Robert Pool is a Washington, DC-based freelance science writer and a contributing correspondent for *Science* magazine. His publications have appeared in a number of places, including *Discover*, the most widely read general science magazine in the United States, and *The Washington Post*. He worked as News Editor at *Nature*, the prestigious London-based international science journal, before leaving to devote all his time to science writing.

THE NEW SEXUAL REVOLUTION

Robert Pool

Hodder & Stoughton
LONDON SYDNEY AUCKLAND

Permissions

Chapter 2
Janet Lever: © 1975 The Society for the Study of Social Problems
Dianne McGuinness: © 1990 JAI Press

Chapter 3
Jared Diamond: © 1993 Discover Magazine

Chapter 5
Stephen Jay Gould: Reprinted from The Mismeasure of Man by Stephen Jay Gould, with the permission of W W Norton and Company, Inc Copyright © 1981 by Stephen J. Gould

Chapter 11
Doreen Kimura: © 1992 Scientific American

Copyright © 1993 by Robert Pool

The right of Robert Pool to be identified as the Author of the Work has been asserted by him in accordance with the Copyright, Designs and Patents Act 1988.

First published in 1994 by Hodder and Stoughton.
A division of Hodder Headline PLC

10 9 8 7 6 5 4 3 2 1

All rights reserved. No part of this publication may be reproduced, stored in a retrieval system, or transmitted, in any form or by any means without the prior written permission of the publisher, nor be otherwise circulated in any form of binding or cover other than that in which it is published and without a similar condition being imposed on the subsequent purchaser.

British Library Cataloguing in Publication Data

Pool, Robert
New Sexual Revolution
I. Title
306.7

ISBN 0-340-58440-8

Typeset by Keyboard Services, Luton

Printed and bound in Great Britain by
Mackays of Chatham plc, Chatham, Kent

Hodder and Stoughton Ltd
A division of Hodder Headline PLC
47 Bedford Square
London WC1B 3DP

To the love of my life,
My sage and my wife,
Amy

CONTENTS

Acknowledgements	ix
Introduction: Hormones and Heroines	1
Chapter 1: Different but Equal	13
Chapter 2: A Tale of Two Sexes	37
Chapter 3: Beyond the Birds and the Bees	67
Chapter 4: Echoes of the Womb	85
Chapter 5: My Brain's Bigger Than Your Brain	113
Chapter 6: Not Quite the Opposite Sex	137
Chapter 7: Variations on a Theme	157
Chapter 8: Raging Hormones	181
Chapter 9: Nature/Nurture	203
Chapter 10: Echoes of the Past	235
Chapter 11: Where Do We Go From Here?	249
Notes	273

ACKNOWLEDGEMENTS

Most special thanks go to Janice Juraska and Sheri Berenbaum, both of whom not only spent many hours describing their research to me but also read through drafts of the entire book, catching mistakes and making suggestions for improvements. I'm particularly indebted to them.

Besides these two, many other scientists took time both to speak with me and to review one or more chapters: Laura Allen, Michael Bailey, Camilla Benbow, Thomas Bever, Mickey Diamond, Laurence Frank, Steve Gaulin, Brian Gladue, Roger Gorski, Elizabeth Hampson, Marie-Christine de Lacoste, Simon LeVay, David Lubinski, Diane McGuinness, Cecile Naylor, Dominique Toran-Allerand, Christina Williams and Sandra Witelson. The book would have been a much poorer one without the help of all these people. Additionally, dozens of scientists and others familiar with the field provided information and descriptions of their work. These included: Gerianne Alexander, Nancy Andreasen, David Ankney, Phil Bryden, Alison Carlson, Victor Denenberg, Richard Doty, Adam Drewnowski, Alice Eagly, Jo-Anne Finegan, Richard Green, Melissa Hines, Michel Hofman, Julianne Imperato-McGinley, Douglas Jackson, Arthur Jensen, Andrew Kertesz, Doreen Kimura, Bruce McEwen, Jeannette McGlone, Krista Phillips, Barbara Sherwin, Irwin Silverman, Rita Simon, Jane Stewart, Richard Whalen, Lee Willerman, James Wilson, Pauline Yahr and Ken Zucker. I'm also grateful to Diane Halpern for her wonderful book, *Sex Differences in Cognitive Abilities*, which is the most complete and objective summary available on what scientists know about differences in intellectual abilities between men and women. I used it extensively as a reference work and as a guide to the most recent research in the field. For anyone who finishes this book and wants to learn more about sex differences, her book is the place to go.

I thank Tancy Holden for helping get me started on this subject, Carol Putnam for reading a draft of the book and offering suggestions, John

Brockman for ensuring the book got published, and Erica Marcus for all her help with the manuscript.

My biggest thanks go to my wife, Amy. Not only has she been good-natured about serving as an example throughout the book, but she has helped shape the book through many discussions with me and with her thoughtful criticism and advice. She read early drafts of all the chapters, telling me what could be made better and, in one case, convincing me that an entire chapter needed to be redone. Indeed, she was so much help that I often threatened to list her as co-author. Having decided to spare her that fate, however, I can only note that it would have been a much different and inferior book had I been working on it alone.

INTRODUCTION

Hormones and Heroines

Across the plains of Africa roams one of the world's strangest creatures. In a violent world populated by such ill-tempered nasties as lions, rhinos and cape buffalo, the spotted hyena stands out for its ferocity. No bigger than a large dog, a single hyena will tackle a wildebeest five times its size, and in a pack it hunts animals as large as full-grown zebras. Once the prey is down, the hyenas devour it quickly in a shark-like feeding frenzy. A pack can polish off a zebra, bones and all, in half an hour, leaving only a few scraps behind.

But it is not the hyena's ferocity that makes it so strange, nor even the demented cackle it uses to communicate and which has earned it the nickname 'laughing hyena'. Look closer, and you'll see what I mean.

The snarling, snapping creatures gorging themselves on zebra meat look much like round-eared, bushy-tailed dogs. They have strong jaws, dirty yellow coats with dark brown spots, and Mohawk-style manes running down the backs of their necks. Their forelegs are slightly longer than their hind legs, which makes them seem slightly unbalanced. And hanging down between the back legs of each hyena is a long, tubular organ. That's right – every member of the pack, from young pups to old pugs, fathers, mothers and godparents, displays what is unmistakably a penis.

It is an unsettling sight – and one that has given rise to endless speculation and legend over the centuries. Some observers have concluded that the species has no females, while others have thought that the animals are all hermaphrodites – creatures with both male and female sexual organs. As late as the 1960s, one South African researcher reported claims that the hyena could change sex, first fathering litters as a male and then giving birth to them as a female.

The truth is stranger than any of these fictions, and it offers a rare glimpse into the inner workings of sex. Ironically, the seemingly unisex hyena illustrates better than any other animal just what it is that makes males and females different in any species, including humans.

A pack of spotted hyenas does indeed include both sexes, and the larger and more aggressive members are the females, each of which sports a penis. Well, it's not *literally* a penis – it's merely a clitoris so enlarged that it looks like a penis – but in size, shape and appearance it's hard to distinguish from the male's sex organ. In addition, the female's labia are fused to form a 'scrotum' which contains fat and connective tissue that give the semblance of testes. As a result, the male and female genitals are so similar that until very recently even experienced field observers had trouble telling the sexes apart.

The female spotted hyena even has erections. In social interactions, two hyenas – male or female – will approach one another, stand head to toe with a rear leg raised, and inspect each other's erect penises.

As you might imagine, hyena sex is a little, umm . . . different. Stephen Glickman and Laurence Frank, two scientists who have raised a colony of spotted hyenas on a reserve near the University of California at Berkeley, have watched it many times but say they have never lost their wonder. For intercourse, Frank says, the female retracts her clitoris/penis and draws it into her abdomen, inverting it 'just like when you reach down into a sock and turn it inside out'. There are special retractor muscles running along her penis just for this job. The inverted female penis forms a pseudo-vagina into which a male hyena must maneuver his own penis, but since only a small hole is left in the female's abdomen after the retraction, this maneuver is something that takes practice. Glickman and Frank report that young males are pretty clumsy the first time they try it.

When a female gives birth the babies come out through the penis, which must stretch greatly to accommodate them. This doesn't seem to be a particularly efficient way of delivery, the two researchers note (in fact it sounds downright painful to most men who hear this story), and an unusually high percentage of pups are born dead or die shortly after birth, apparently due to birth complications.

This is all so bizarre that the Berkeley duo and other scientists are still trying to figure out exactly what happened to the spotted hyena and why. The bottom line seems to be that nature has turned the masculinity knob all the way up for these beasts.

In every mammal, including man, maleness or femaleness is

determined in the womb according to the sex hormones that are present during development. High levels of male hormones set a fetus along the path to becoming a male, and the fetus will become female if these male hormones are mostly absent. (Later in life sex hormones continue to influence both body and brain, although the effects are not so major.) Normally it is the fetus itself that provides most of the hormones in the womb, determining its own sex according to its chromosomes, but in the case of the spotted hyena the mother takes a hand. The blood of a female hyena contains a tremendous amount of a hormone called androstenedione, and during pregnancy the placenta converts this chemical into testosterone, so that both male and female fetuses are bathed in male hormones.

One result of this testosterone bath is that the female hyena's external sex organs take on a masculine shape, but the effects don't stop there. After developing in this super-masculine fetal environment, spotted hyenas emerge from the womb as the most aggressive infant mammals known to science.

Equipped with a full set of incisors and canine teeth, newborn spotted hyenas may face a fight to the death before they've even had a chance to nurse. Cubs usually arrive as twins or triplets, born an hour or so apart, and as soon as the second one arrives, the first immediately attacks it, biting it on its neck and shaking it. The younger one counterattacks, and over the next few hours or days the better fighter establishes dominance. If the twins are of the same sex, the dominant one usually prevents its weaker sibling from nursing, slowly murdering it with a combination of malnutrition and infected wounds. If the twins are different sexes, the dominant one may let the other survive, but the 'top dog' nurses more and becomes larger, an advantage that it keeps into adulthood. The mother cannot intervene in the fighting because the birthing dens are too small for her to enter – typically she gives birth at the mouth of an abandoned aardvark tunnel and the newborns immediately crawl into it, where they are protected from predators but not from each other.

It's hard to understand why any animal should be so aggressive that it kills its own siblings, but Frank and Glickman have offered one suggestion. A pack of spotted hyenas is particularly nasty at mealtime, which resembles a cross between a crowded buffet and a free-for-all riot, and the two researchers have hypothesized that females had to become aggressive in order to compete for food and provide enough for their cubs to survive. One way to do this was for the females to become more male-like, and eventually the female hyena got even bigger and meaner than the male. In the hierarchy of a hyena pack, each female is dominant over every male, and the pack's dominant member is invariably female.

The hyena's sexual equipment is unique. Glickman and Frank say they know of only one other mammal, a species of mole, whose females have a clitoris blown up to the size of a penis, but the female moles at least have a separate vagina which they use for sex and to deliver babies. The spotted hyena is also unique in its aggressiveness – no other mammal drops out of the womb ready to fight and bite.

This uniqueness makes the hyena a spectacular example of the power that sex hormones can exert over the development of body and brain, but that power is by no means unique to the spotted hyena. Every mammal, from mole to man, is shaped by sex hormones in the womb. All that sets the hyena apart from other animals is the exaggeratedly high level of hormones and the fact that both sexes receive a jolt of testosterone in the womb, not just one.

In other mammals, where only the male sees high levels of male hormones in the womb, the two sexes develop distinct sets of sexual organs. Furthermore, the brains of the two sexes are wired differently, resulting in different types of behaviors for males and females. Scientists know of many examples. A lab rat's sex life, for instance, is completely determined by sex hormones – given male hormones, a rat will act like a male and mount female rats, and otherwise it acts like a female and responds sexually to males. If the rat is artificially exposed to the wrong sex hormones while its brain is developing, it will be sexually confused for the rest of its life. Sex hormones also influence such sex-role behaviors as a mother nurturing a child or two males fighting for dominance, as well as some behaviors that have no obvious connection with sex, such as how a rat finds its way through a maze.

What about humans? That's the sixty-four-thousand-dollar question. We know that whether a person has a vagina or a penis, breasts or a beard depends on the sex hormones, but what about the mind? Boys and girls, men and women certainly behave differently – males are more aggressive, for instance, and females more nurturing – but how much of that can be traced to the same type of hormone action on the brain that makes spotted hyenas, both male and female, such combative creatures?

The answer has been slowly taking shape. It is not an easy question since human brains are much more complex than those of rats or hyenas and much of human behavior is learned. If a woman is more nurturing than a man, is she heeding her hormones or responding to everything she has been taught since she was a small child? If a man is aggressive, is it testosterone or too much pro hockey on television?

Complicating the issue even more is the fact that it is intensely politicized. As women's roles have changed and more and more women have gotten educations, pursued careers and fought their way into previously all-male bastions, our society has been forced to rethink the traditional roles of the sexes. Are women just as capable, on average, of doing any and every type of job as men are? Can men do just as good a job of raising children as women can? Many feminists, with their goal of complete equality between the sexes, have insisted that there are no innate mental or psychological differences between men and women – only differences produced by society – and they have denounced any evidence to the contrary as 'sexist'. Some have gone so far as to say that research into biological differences should not be done because of its potential for misuse, and even those who accept the existence of biological differences between the sexes worry that too much publicity about them could lead to discrimination against women. If, for example, girls were shown to have on average a slightly lower aptitude for math than boys, would society stop encouraging girls to pursue careers in science or use this disparity as an excuse for paying women less than men?

In part because of these worries, much of the research into sex differences over the past two decades has been devoted to showing how societal influences are to blame for many of the dissimilarities between men and women – how upbringing, peer influence and social pressures push males and females in different directions. That research has confirmed that boys and girls indeed are treated differently from the time they are born, and that the effect of this differential treatment is often to encourage children to behave in ways that are stereotypical for their sex. Although this may seem obvious in retrospect, at the time it was vitally important to focus attention on how much we can affect boys' and girls' lives with our teachings and expectations. Twenty or thirty years ago, for example, no one gave much thought to the damage that parents and teachers can do to their daughters and female students by telling them 'girls can't do math'. Today, thanks to the consciousness-raising that followed the research socialization, fewer people make such mistakes.

But a growing number of researchers have become convinced that environment cannot explain all of the differences between the sexes and that despite the difficulty and the controversy of the work, the path to understanding these differences passes through the womb. 'The brain is the frontier now,' says Sheri Berenbaum, a psychologist at the Chicago Medical School. 'When I started graduate school [in the early 1970s], I really believed in the social environment,' but since then, 'the data have convinced me that biology plays a big role.' So Berenbaum has spent the past decade studying how children's behavior is influenced by the levels of sex hormones they were exposed to in the womb.

The picture of humanity that Berenbaum and others have been piecing together is one that accepts society's influence but which insists that sex differences begin long before the newborn is taken home from the hospital and placed in the color-coded baby's room, blue for boys and pink for girls. The differences do not imply that one sex or the other is superior – instead the scientists see men and women as inherently equal, often complementary versions of humanity – but many of these differences will not go away, no matter how much we wish they would. The changes produced by the sex hormones in the womb echo through our lives, diminished and distorted by the intervening years, but never quite fading away.

The scientists who do this research are a rather independent-minded, thick-skinned bunch. They've had to be. They're working in an exceptionally sensitive area and they often find their work closely scrutinized by other scientists, feminists and the media. Criticism and complaints can be loud, pointed and sometimes personal, and more than one scientist has left the field for one with a lower profile.

Who are these mavericks? Many of them are young, having started college in the late 1960s or 1970s, so that they have been Ph.D. scientists for less than ten or fifteen years, and some of them for less than five. A majority of them are psychologists of one sort or another, although they are joined by researchers from a spectrum of other areas: endocrinology, neuroanatomy and behavioral ecology, to name just three. And after you've read their research reports and talked to the scientists themselves, one truly striking feature emerges about this sex difference fraternity: it's really a sorority. Most of the scientists doing the provocative, ground-breaking research into human sex differences are women.

That surprises a lot of people. Many feminists, including some feminist scientists, are suspicious of and even hostile toward this research. They argue that if women are shown to be innately different from men, then that information will be used as a rationale for keeping women in the kitchen and out of the boardroom. It's only a short step, they say, from 'different' to 'inferior'. The feminist biologist Anne Fausto-Sterling, in her influential book *Myths of Gender*, claimed that many of the scientists finding sex differences really want to suppress women, although they might not be consciously aware of it: 'More often than not their hidden agendas, non-conscious and thus unarticulated, bear strong resemblances to broader social agendas.'

The reality is quite different. Not only are many of the researchers into sex differences women, but almost all of the women scientists in this field identify themselves as feminists or at least sympathize with many feminist goals. Most of them will tell you that they themselves have experienced sex discrimination of one sort or another during their education and careers, and each has thought deeply about the possibility that her work could be used as justification for discrimination against other women. They are not fools or tools of male-dominated society, nor do they have any hidden agendas, and they uniformly resent such implications.

So why is it mostly women who are exploring human sex differences today? The researchers themselves wonder about it sometimes and will gladly speculate over coffee, but nobody really has a good answer. In researching this book, I found that if I asked a female scientist why she studies sex differences, her answer would be much the same as a male colleague's. She would talk of the intellectual challenge or the satisfaction that comes from learning something that no one has ever known before. But by pushing for more specific answers – 'Why did you choose sex differences in particular?' – and speaking to many of these women, I did begin to see a pattern emerge.

Some female scientists say that they view the existence of sex differences – particularly those that favor males – as a personal challenge. Camilla Benbow at Iowa State University is one of them. She began her career studying seventh- and eighth-grade children who score exceptionally high on tests of mathematical ability, and at first her goal was simply to help these 'mathematically precocious youth' turn their talent into adult success. She got sidetracked, however, by one fact that both intrigued and bothered her: among the very best math students, boys outnumber girls by as much as thirteen to one. 'If you're a female,' she says, 'it's hard to understand why it is that males score so much better.' That question has driven her over the past dozen years to become the United States' leading expert on mathematically talented children and why the boys and girls among them end up pursuing very different careers.

Other women researchers say they got started in sex differences because male scientists were slighting the female half of psychology. Take Janice Juraska, who studies sex differences in rats at the University of Illinois. She began college in the late 1960s as a chemistry major, but 'it seemed very irrelevant at the time,' she remembers. 'I declared psychology almost as a whim. I had a hard time identifying with people who cared more for chemical reactions than human reactions.' As a graduate student she studied how environment affects brain structure in rats, but it bothered her that the experiments were traditionally done only on male rats. 'No one seemed concerned about whether the same was true for female brains,' she says, and when she asked why not, her predominantly male colleagues had two standard responses: female rats are too messy to study because their hormone levels fluctuate so much, or else, 'Of course females are the same as males. Why worry about it?' Juraska did worry about it, and in the process has discovered some startling differences in how the environment affects the brains of male and female rats.

In short, sex differences seems to be a much more personal issue for women than for men. More than one female scientist suggested to me that the reason so few men enter the field is that 'men don't feel different, women do'.

Still, this is only part of the story. Many of the women doing sex difference research – perhaps a majority of them – will tell you that sex isn't really their main focus at all. They choose to study sex differences not out of curiosity about why men and women are different but because sex differences offer a powerful tool for studying the mind in general. Christina Williams at Barnard College of Columbia University, who showed that sex hormones influence how a rat learns a maze, is typical. 'I don't think of myself as someone who is interested only in sex differences,' she says. 'I'm interested in [brain] development.' Working with sex differences gives her an alternative to the standard laboratory technique for learning how the brain works, which boils down to destroying part of the brain and seeing what goes wrong. 'I don't like looking at damaged systems,' Williams says, and by focusing on ways in which male and female brains do things a little differently, she has 'two perfect brains to compare' instead of looking at healthy brains versus damaged brains.

But if sex differences offer such a powerful tool for studying the mind, then why don't more male scientists choose to use it? This was the other part of the pattern I discovered in talking to researchers.

I got one clue from Sheri Berenbaum when she told me how she first became interested in sex differences. She was a graduate student at the University of California at Berkeley in the 1970s working as an assistant for a male professor. When it came time for him to give a lecture to his class on sex differences, he asked her to do it for him – even though she had never specifically studied the subject. He thought it would be 'less politically offensive,' Berenbaum said, if a woman were to tell the students how males and females differ.

That male professor at Berkeley was not unique. Most of the men I spoke with agreed that they're very careful what they say in public about sex differences. At one university, a male psychologist pulled out a recent newspaper article on sex differences and showed me a quotation from Canadian psychologist Doreen Kimura, one of the leading sex difference researchers in the world. Kimura had said, 'The idea that men and women should be equally represented in all occupations and all professions is simply ridiculous given the data we have on their different ability patterns.' The male researcher agreed with Kimura, but confided that he would never have said that for publication and certainly not as bluntly as Kimura. A man who said that would be attacked as a sexist, he thought. (To be fair, the women in this field also watch their public words very carefully, but it's more from a worry that their comments might be misinterpreted than a fear of being attacked as sexist. Men have both concerns to deal with.)

Sex differences in animals are not nearly so controversial as in humans, and male researchers traditionally have done much of the work on rats, monkeys and other lab animals, but even in these fields men know they must watch what they say. Bruce McEwen at Rockefeller University in New York City is a leading expert on how sex hormones direct activity in the rat brain, and he is often asked by the press for comments on human sex differences. 'There have been occasions,' he says, 'where I've just said, "I'm not going to answer that. Ask Doreen Kimura."'

At the University of California, Los Angeles, Roger Gorski is another authority on sex differences in the rat brain. In the late 1970s he discovered a part of the rat brain that is five times larger in males than females – still the record for largest structural sex difference in a mammalian brain. But when a graduate student, Laura Allen, told him she would like to look for similar differences in the human brain, he discouraged her. It was too controversial an area, he said, and she should stick with the rat brain. 'I was worried about the sensationalism,' he remembers. But Allen persevered, convinced that sex differences in brain structure might hold clues to such things as sexual orientation – why most men are sexually attracted to women and vice versa. Now Allen, just a few years after earning her Ph.D., has found more areas of the human brain that differ between males and females than any other researcher.

In a sense, the female dominance in human sex difference research appears to be a rather ironic consequence of the women's movement. Feminists succeeded in making the research controversial enough to keep many male scientists from getting directly involved, but at the same time the emphasis on bringing females into traditionally male fields led more and more women to enter scientific research. It is these women, sometimes against the advice of male advisers, who have done so much of the work in tracking down sex differences.

It is a heroic quest, as the female scientists in this field often find themselves in conflict with other women and sometimes even with themselves. 'There is a lot of feminist thought that is quite hostile to what I do,' Juraska says, 'and I've tried to change that at the local level.' She has spoken to women's groups to describe her work and its implications and to explain why it should not be seen as threatening to women, and she believes that she has made some headway, although there are still women who 'really don't want to hear it'. But before Juraska reached this point, she herself had to decide how she felt about the subject. 'This has been all too personal. Growing up as a smart female I got mixed messages about a woman's role,' and when she began to study sex differences it forced her once again to 'wrestle with the question of what my femaleness fated me to. Some people might stay away from the area, but I chose to go after it.'

The quest that Juraska and others have embarked on has yielded some provocative details about how and why the sexes differ. We now know, for instance, that hormones in the womb do influence behavior later in life and that girls who are exposed to high levels of male hormones in the womb will act, play and think more like boys. Researchers have shown that men's and women's brains are structurally different and that the sexes use different parts of the brain to solve certain types of problems. These differences in the brain may be responsible for many of the dissimilarities that psychologists have found in how men and women approach the world. Women are more interested in people and men are drawn more to objects, women rely more on words and men are more likely to visualize a problem, and women focus on the give and take of relationships while men pay attention to who is dominant over whom.

But more significant than any of the individual details is the overall vision of man and woman that is emerging from this research. No longer can all the differences between the sexes be written off as caused by socialization, but neither can people continue to hold to the old stereotypes of manly men and womanly women. Although research has shown that certain characteristics can indeed be considered 'masculine' and others 'feminine', the same research has shown that none of these characteristics are found exclusively in one sex or the other. Most people are a blend of both male and female traits, and it's perfectly natural and normal, for example, if a man is nurturing or a woman is aggressive. The most important lesson of the research is that none of the differences imply one sex or the other is superior. Men and women are equal partners in humanity.

In many ways the emerging understanding about the differences between the sexes couldn't have come at a better time. Over the past thirty years, the traditional Western ideas about the proper roles of men and women have broken down, with no consensus in sight on what should replace them. Should men and women be equally willing to stay home with the children and put their careers on hold? Should there be equal numbers of men and women in every occupation? Schools have traditionally offered remedial reading programs whose students are mostly boys – should there also be remedial math programs in junior high and high school to help girls keep up with boys in mathematics? Or, on a more down-to-earth level, should parents encourage little boys to play with dolls and little girls to play with trucks?

It would be naive to think that society's answers to these questions will be based purely on what science has to say about sex differences. After all, even researchers who agree about the scientific facts often disagree sharply about their implications for social policy. On one trip to research this book I visited a female researcher who argued vehemently against any type of quotas or preferential hiring aimed at increasing the number of women in mathematics or engineering fields, pointing out that males seem to be naturally more interested in such work and somewhat better at it. The best solution, she said, would be to make sure that everyone has the same chance to succeed, leave it at that, and accept that there will probably always be more male than female mathematicians. Two days later, I had lunch with a second woman scientist who agreed with the first that males do seem to have some biological advantage in math-related areas, but who then argued that society should therefore intervene to guarantee equality between the sexes in such jobs.

Nevertheless, the history of science shows that ideas have consequences and that a scientific consensus usually trickles down to the general population. Science may not determine popular beliefs – if it did, no one would read horoscopes – but it does wield a certain constrained power over them. Almost nobody these days thinks that the sun moves around the earth.

In that sense, the research on sex differences is some of the most meaningful science being done today. It not only has implications for broad social issues such as educational policy or family leave practices, but more importantly it has the potential to change how people view themselves and others. The question of 'What is man?' or 'What is woman?' is central to each person's sense of identity and to how each person defines his or her place in the family, the workplace, the social circle and in humanity, and it is those questions that lie at the heart of the research described in this book.

CHAPTER 1

Different but Equal

It was still dark outside as Camilla Benbow, seven months pregnant, headed downstairs to wake her firstborn and get him ready for school. Stepping past the rooms of three-year-old Bronwen and one-year-old Trefor, she walked into Wystan's room. Wystan was five, and he didn't like getting up this early, but Mommy had little sympathy. A twenty-four-year-old graduate student with a husband, three and a half children and a doctoral dissertation to complete, she knew all about wanting to stay in bed for just a few more minutes, but there never seemed any time. Gently she pulled her son out of the covers and helped him pick out some warm clothes. It was the fifth of December in 1980, and Baltimore was getting cold.

Upstairs her husband, Robert, was in their bedroom getting dressed and listening to the news on National Public Radio. Suddenly he shouted something down to Camilla and turned up the radio loud enough that she could hear it all the way in Wystan's room. Today's headline story, it seemed, was her.

'I remember hearing it and just not believing it,' Benbow says now. She knew that her research, which showed that junior high school boys had more mathematical ability than girls of the same age, would probably start some fights in the scientific community, but she never dreamed how much attention it was about to get from the popular press. It had started the night before, when she got calls at home from *Time* and *Newsweek* reporters who had heard about her research, due to be published the following week, and who didn't want to wait until the next day to talk to her. She answered their questions from her kitchen without realizing that they were just the first drops in what was about to become a downpour.

It finally began to sink in the next morning after she had packed Wystan off to school, grabbed a bite to eat and headed for her office. As soon as she walked in, about a quarter after eight, a wide-eyed secretary handed her a stack of messages. 'I still remember the look on her face.' Several newspapers and radio stations had called for her in just the past fifteen minutes, and the phone was ringing again. It never stopped. 'I spent the whole day talking to reporters. I would get off the phone with one and another call would come in.' Later that morning, when the local television stations phoned to say they were sending camera crews out to interview her, she looked down at the plain maternity dress she was wearing and wished that she had planned things better. 'I just put on my regular stuff to go to the office.' As a graduate student and an expectant mom, she didn't have a lot of clothes to choose from, 'but I did have some nicer things to wear. I just wasn't expecting anything.' When she thought about how those clothes would look on the TV news, 'I considered going home to change, but there was no time.'

She finally escaped the office around five thirty, in time to get home for the local evening news. She and Robert gathered the children around the small black and white television set in the living room, and they all sat down to watch Mommy's TV début. 'The kids thought it was pretty neat,' she says. None of them minded that she hadn't gotten dressed up.

The research that had caught the eye of reporters around the country was a provocative and troubling scientific study Benbow had performed at Johns Hopkins University under the guidance of Julian Stanley. For eight years Stanley, a well-known educational researcher, had been collecting information on a group of super-smart seventh- and eighth-graders, boys and girls who had scored in the top two to five percent on one of the standardized math tests given to American school children. Over the years, Stanley had identified about 10,000 of these mathematical whiz kids (nearly 5,700 boys and almost 4,300 girls) and subjected them to the SAT – the dreaded Scholastic Aptitude Test used by many colleges and universities to weed out applicants. Even though they had never taken high school courses, these 'mathematically precocious youth', as Stanley called them, scored as well on the SAT as the average high school junior or senior.

Stanley had started the program to find mathematically bright children early and help them develop their talents, but Benbow saw another use for it. She could test one of the enduring sex stereotypes: that boys are better at math than girls.

When Benbow analyzed the SAT scores of the mathematically precocious youngsters by sex, she found a striking pattern. Even though the male and female averages on the verbal portion of the test were nearly identical, the boys averaged thirty to thirty-five points more than the girls on the mathematical section. That difference was dramatic enough, but when Benbow zeroed in on the smartest of these very smart kids, she discovered an even more extraordinary disparity. Among those students who scored above five hundred points on the math section – which is an average score for a male, college-bound high school senior but very good for a seventh- or eighth-grader – Benbow found twice as many boys as girls. Above six hundred, boys were four times more numerous than girls. And among those who scored over seven hundred – the truly brilliant group most likely to have the innate talent to become mathematicians or theoretical scientists – Benbow found that boys outnumbered the girls by thirteen to one.

Those numbers grabbed the attention of the media and the public, but it was another part of Benbow's paper that stirred up the scientific community. The facts were, after all, indisputable – Benbow had been meticulous in her data gathering and analysis – and although some of her peers might think it was bad form to publicize the male mathematical advantage, none could challenge it. But Benbow had ventured past the numbers and into the mine-field of possible causes.

Indeed, the whole point of Benbow's work was to examine the reasons for the male advantage on standardized math tests, a phenomenon that was already well known, even though she was one of the first to study it in junior high school students. Since at least the late 1960s male high school seniors have scored an average of forty-five to fifty points higher than their female peers on the math section of the SAT. (That's about eight percent of the total possible points on the test, which is scored from a low of two hundred to a perfect eight hundred.) But many researchers thought there was a simple explanation for this difference that had nothing to do with biology: high school boys take more math courses than high school girls. No need to invoke genes or hormones – it was simply a matter of boys learning more mathematical concepts and getting more practice solving math problems.

Benbow's research shattered that theory. Her numbers showed that most of the sex difference in SAT math scores was already present by the seventh grade – when boys and girls have taken the same math courses.

As she followed the mathematically precocious youth through high school, she saw that the boys in the study increased their advantage over the girls to an average of fifty points by their senior years, but she saw no evidence that the change was due to differences in courses taken. The mathematically precocious boys and girls took the same number of math courses until twelfth grade and girls had slightly better grades.

Benbow went one step further and evaluated two other social hypotheses for the female underperformance on standardized math tests. Some researchers had blamed it on girls being taught to dislike mathematics by parents and peers who tell them it's a male subject, while others suggested that girls often lose interest in math because they don't plan to be scientists or engineers when they grow up and thus don't think it will be of use to them. Benbow asked the boys and girls in her study how much they liked math and how important they thought it was for their future careers. She found no relation between either factor and how well the students did on the SAT.

Furthermore, by giving the SAT to seventh- and eighth-graders, who hadn't taken the courses needed to learn how to work the problems, Benbow and Stanley believed they got a purer measure of mathematical ability. The scores of the young students would depend less on how well they learned material in class and more on how quickly they could figure out how to solve a strange new problem.

All of these factors, Benbow and Stanley said, imply 'that sex differences in achievement in and attitude toward mathematics result from superior male mathematical ability.' They acknowledged that they hadn't ruled out every possible social explanation for the sex differences, but they insisted that biology seemed to be responsible for at least part of the male superiority.

That paper, which appeared in December 1980 and was still being written about months later in newspapers and magazines around the country, marked a major shift in the study of sex differences. Throughout the 1970s, much of the research done on how the sexes differ was based on the assumption that the disparities are mostly caused by differences in how boys and girls are raised and in how they're treated by society. It's easy to find such differences: boys are encouraged more by their parents to play with blocks and other toys that help them develop spatial ability, which is important in math, girls get less attention and encouragement in classrooms, and so on. What's not easy is to test just what effect these differences in treatment actually have. If parents were to give their little girls just as much encouragement to play with blocks as little boys get, would the girls' math scores fifteen years later be any better? It's hard to know.

By investigating widely accepted social explanations for the sex difference in math, showing they didn't work, and concluding that biology probably does play a role, Benbow was reflecting a new approach to sex difference research that has gradually come to the fore over the past decade. More and more scientists have begun to accept the possibility that differences in behavior and abilities between males and females are due, at least in part, to physical differences in their brains. Although Benbow's paper was not responsible for this change, it did kick off a series of provocative and highly publicized studies by young, imaginative, mostly female researchers who have put the biology back in sex differences. It was almost as if a gauntlet had been flung down, signaling a challenge to the accepted environmental dogma.

The flinger of that gauntlet, when first you meet her, seems an unlikely person to inaugurate such a fight. Soft-spoken and still carrying a bit of a lilt from living her first nine years in Sweden, Camilla Persson Benbow calls to mind a shy Scandinavian country girl with her curly brown hair, light eyes, strong hands and fair skin. But talk with her a while, or look more deeply into her eyes, and you start to understand why the Swedes ruled the northern seas for so many centuries. There is a strong, unflappable core here, and it takes little imagination to see her as some Viking warrior's wife of ten centuries ago, born to raise fierce warrior sons and rule over the homestead until her husband returns from raiding Scotland or Germany – and, if truth be told, even after he returned. In these more civilized times, when brawn and swordsmanship are out and brains are in, she herself leads the raids on rival strongholds.

Ironically for someone who so challenged feminist social theories of sex differences, Benbow seems the epitome of the 'superwoman' ideal of 1970s feminism, merging a highly successful career with motherhood. At eighteen she married Robert Benbow, a molecular biologist twelve years older whom she had met when her father worked at Caltech. While Robert held jobs first at Cambridge University in England and then at Johns Hopkins, Camilla studied psychology. She earned her bachelor's degree by the time she was twenty, having taken only a short break the year before to give birth to Wystan (a traditional Welsh name). Six more children would follow at approximately two-year intervals, during which time she would earn her doctorate, publish a large amount of original and highly respected research, and become chairman of her department in her mid-thirties.

As an undergraduate Benbow had thought she would become a clinical psychologist, but after taking a course from Julian Stanley in her senior year, she found herself intrigued by the mathematically precocious kids and joined his team of research psychologists when she entered graduate school. Even before Benbow came aboard, it had been no secret among Stanley's group that junior high school boys did much better than girls in math. As early as 1972, in the first group of seventh- and eighth-graders that Stanley tested, nearly twenty percent of the boys scored above six hundred on the math part of the SAT, but no girls did, and the pattern persisted year after year. Yet few of Stanley's students before Benbow had thought the sex difference was important enough to pursue as a separate issue. Benbow did. She told Stanley she'd like to test whether the sex difference could be linked to such factors as boys taking more math courses than girls or liking math more, and then eventually publish a paper on it. He immediately agreed, she remembers. He'd thought for some time it was important to report the data, but he hadn't made it as high a priority as Benbow would. 'Women scientists may be more interested in the subject,' Benbow suggests, 'because they're on the wrong side of the gender gap.'

When her paper was published in the prestigious journal *Science*, Benbow found herself in a spotlight hotter than she had expected. She could read about her study in just about any newspaper in the country, but in many of those articles there were two or three other scientists quoted who didn't like her conclusions. 'If your mother hates math and your father tells you not to worry your pretty little head about it, do you think that a math test would be an accurate measurement of your ability?' said one. 'I think they are on darned shaky ground when they draw conclusions about genetic differences,' said another. Some got quite personal in their attacks: 'I think that what we have here are people who are looking for a justification of their own political beliefs,' one biologist told a Canadian newspaper.

The sharpest complaints came from those who worried that Benbow's data would hurt females' chances in math and science – that some people might use her studies as a rationalization for policies that discriminate against women. 'It is a political issue,' said Sheila Tobias, feminist and author of *Overcoming Math Anxiety*, a book about girls and math ability. 'This will influence the behavior of the educational community ... I am in this research [on girls and mathematics] to get at the truth, but also for a political motive. I want to get the country to accept this premise: that potentially girls and boys can perform equally.' And a month after Benbow's article appeared in *Science*, that journal gave equal time to Wellesley College mathematician Alice T. Schafer and American University mathematician Mary W. Gray. In an editorial response the two women argued that 'environmental and cultural factors have not been ruled out,' and warned that press reports of Benbow's work could hurt efforts to get more women into mathematics. 'It is virtually impossible to undo the harm that the sensationalized coverage has done,' they wrote.

'It was tough to see some of the stinging criticism that came out, much of which was not accurate or fair,' Benbow says, 'but I just dug in my heels. My reaction was, "I'm going to show you."' She was encouraged because none of her critics had denied the accuracy of her data and because they quietly dropped the argument that taking more math courses was the cause of the male superiority. Instead they fell back on 'subtle environmental influences' that were much harder to pin down, Benbow says.

So Benbow has spent the past decade testing every subtle environmental influence she could think of that might explain the male math advantage among mathematically precocious youth. At first working with Stanley and then taking control of the study herself, she has accumulated SAT scores from more than a million seventh- and eighth-graders and questioned many of those students on their family background, medical history, personal likes and dislikes, career plans and so on. So far: nothing. Mathematically precocious girls have no more math anxiety than their male peers. In fact, these girls like math. They think math will be helpful to them in future careers. They get just as much encouragement and help from their parents as boys do. And, if anything, they get even better grades than boys in math courses (perhaps, some researchers think, because girls are more concerned than boys with good grades and living up to the expectations of parents and teachers). Benbow went so far as to check out which toys the students played with as children, to see if boys might get an advantage from playing with blocks or other toys that develop spatial ability. There was very little difference in the types of toys these bright boys and girls played with when young, Benbow says, and those preferences had no relationship to differences in math test scores later on.

'After fifteen years looking for an environmental explanation and getting zero results, I gave up,' she told one reporter. If there were some difference in the way boys and girls were raised that is strong enough to cause such a consistent male advantage in math, she reasoned, it should have shown up somewhere in their attitudes toward math or their performance in math classes. It doesn't, and she finds herself where she started: suggesting that biology must play a role in the male dominance at the top end of mathematical ability.

Benbow, who now chairs the psychology department at Iowa State University, is careful to say that her findings apply only to those students in the top few percent of mathematical ability since those are the only ones she has studied. Her findings may have implications for the recruitment and training of mathematicians, scientists and engineers because those careers demand people whose math skills are superior, but her work doesn't necessarily have any bearing on average kids.

Among those middle-of-the-road children boys have the same fifty-point advantage on the math portion of the SAT, but it is much harder to say what's causing it. It's possible, for instance, that girls have actually been gaining on boys in math but that the gain has been hidden by a change in the students taking the SAT. In the late 1960s, the average score of female high school seniors on the verbal portion of the SAT was several points higher than the male average, but the boys have long since caught up and passed the girls and now they're about ten points ahead. Why? A likely explanation, says Diane Halpern, a psychologist at California State University at San Bernardino, is that the population of students who choose to take the SAT has changed dramatically. As more and more girls have decided to go to college since the 1960s, no longer do only the very top female students take the SAT, and this may have caused the average female score on the SAT to drop. If so, the math sex difference may have been getting smaller even though the average score on the SAT's math portion doesn't show it. (Benbow's data doesn't have the same problem since she has consistently looked only at the top few percent of students.)

A study done by Janet Hyde, Elizabeth Fennema and Susan Lamon indicates that this might be the case. After examining all the studies they could find that tested for a mathematical sex difference, they concluded that the size of the sex difference has been decreasing over the years. If so, some of the male advantage in math twenty-five years ago may well have been due to greater discouragement of females, and as society has accepted that girls can do math, they've closed the gap on boys.

A gap still does remain, however. Hyde, Fennema and Lamon found that although girls are better than boys in computation – addition, subtraction, multiplication and division – and equal to boys in understanding mathematical concepts, the boys gain the advantage in high school when problem-solving skills become important. This pattern of differences makes it difficult to blame the disparities solely on social factors, Halpern points out. If girls have been taught to fear or dislike math, then why do they actually do better than boys in some areas?

We probably won't know for another generation just how much of boys' advantage in math is innate since it takes a long time to erase stereotypes and raise children in a relatively sex-neutral environment, but if the Teen Talk Barbie incident is any indication, we're well on the way to a society that won't stand for even a hint of sexism in math. In the second half of 1992 the giant toy company Mattel brought out Teen Talk Barbie, a version of the classic doll that had been given a voice to go with her blonde hair and 36–18–33 measurements. Mattel had researched the way kids today talk and had come up with two hundred and seventy phrases that a teenager supposedly would utter. Each Barbie was given a random selection of only four of those phrases, thus saving Mattel money and assuring that Lauren's Teen Talk Barbie and Abigail's Teen Talk Barbie wouldn't say exactly the same thing. One of the two hundred and seventy phrases – spoken by approximately one in every sixty-seven Barbies – was, 'Math class is tough!'

Now unless kids have changed a lot since I was young, this is indeed something that normal teenagers – both males and females – do say, except that Mattel probably had to clean up the language a little. But when that doll hit the stores, Mattel found out what 'tough' really means. Newspaper columnists berated the company for propagating negative stereotypes about girls, mothers talked about boycotting Barbie, and outraged citizens wrote letters to the editor demanding that Mattel either take Teen Talk Barbie off the shelves or bring out a Teen Talk Ken who said, 'Barbie, would you help me with my math homework?' Recognizing a public relations disaster when it saw one, Mattel quickly silenced the offending Barbies. Rumor has it that Mattel is now developing a new doll, Blackboard Barbie, dressed in cap and gown and sold with an accessory kit containing a calculator, a piece of chalk, and a blackboard covered with geometric diagrams and algebraic equations. When you pull on her tassel, she says, 'Math class is fun!'

The male advantage in mathematics gets most of the press – and much of the scientific attention – but females have their own areas of excellence. Indeed, for one brief moment at the beginning of the intelligence-testing era, girls came out ahead on the tests, but the rules were quickly changed to make sure that didn't happen again.

It took place some ninety years ago when France's premier psychologist, Alfred Binet, was attempting to develop a simple, objective way to measure intelligence. His immediate goal was to identify mentally deficient children who needed special schooling, but his test would also give a rough ranking of intelligence among normal kids. By 1905 he had settled on a series of tasks and questions that started out easy and got harder and harder, including such things as following simple instructions ('Sit down'), naming an object in a picture, defining common words, being shown a group of objects and then listing them from memory, finding rhymes for given words, and visualizing what a piece of paper would look like after being folded, cut and then unfolded again. How many of these tasks a child could perform provided a measure of 'mental age', and by comparing that with the child's chronological age, Binet had an objective test for whether a child should be considered mentally retarded.

It was the first practical intelligence test and has served as the model for many others. Its descendant, the Stanford–Binet test, is still popular today. This much of Binet's work is well known, but what's not so well known is the rest of the story.

As Binet was testing an early version of this intelligence scale, he found that boys were much more likely to get low scores on it than girls. Uh-oh! Since many scientists of the time thought that females were naturally less intelligent than males – only the most progressive allowed that the sexes might be equal – Binet knew that he must have done something horribly wrong. And it didn't take much looking to figure out what it was. Many of the items on the test were obviously 'biased' in favor of females because the girls were getting them right at a younger average age than the boys. A few items had the opposite bias, with boys answering correctly at a younger age, but there weren't enough of these items to balance out the ones with a female advantage. The solution was obvious: throw out the items with the biggest female bias and add a few that favored the boys. (If Binet threw out all of the items favoring one sex or the other, he'd lose too much of his test.) From then on, Binet was always careful to balance the two types of questions so that males and females performed equally well, and that tradition has been continued in most modern IQ tests.

Psychologist Diane McGuinness tells this Binet story in her book *When*

Children Don't Learn. The standard rationale for the balancing act in IQ tests is that it guarantees the two sexes are treated fairly and equally, but don't be too willing to swallow this line, McGuinness warns. 'If in the early tests more girls had received low scores than boys, it is likely that the test would not have been revised, as it would have confirmed what everyone believed.'

History doesn't say which items Binet removed from his test to make sure that the girls didn't embarrass the boys, but from what we know now we can make a pretty good guess: they probably involved verbal abilities. In general, females have better language skills than boys almost from the time they start to talk. Although there have been conflicting findings about the old stereotype that girls learn to speak earlier than boys, scientists have found that once they do get started, girls are superior in a number of ways.

One example is a study of language development in boys and girls aged two and a half to four done by Dianne Horgan as a graduate student at the University of Michigan at Ann Arbor. To determine the children's grasp of language, she measured their 'mean length of utterance' – the average number of words a child strung together in sentences and other expressions. She found that girls spoke longer utterances at earlier ages than boys, and made fewer language mistakes at the same time. In addition, boys were slower to learn to use such complex constructions as the passive voice – 'The window was broken,' instead of 'I broke the window'. Other researchers have found that female infants babble more than male infants and that young girls use two-word sentences earlier and have larger vocabularies than boys of the same age.

By the time they're in elementary and junior high school, girls have translated their early language advantage into classroom superiority, beating the boys on such skills as spelling, capitalization and punctuation, and comprehension of what they read. A recent cross-cultural study of high school students in the United States and Japan found that girls did much better at remembering details of a story that had been read to them. In short, it's no surprise that Binet's early tests showed an advantage for girls since language skills are the easiest ability to test in young children, and girls have a consistent advantage there.

Strangely, this doesn't translate into a female edge on the verbal section of the SAT, where male high school students average about ten points more than females. Nor is there a female advantage among 'verbally precocious youth' comparable to the male superiority Camilla Benbow found on the math side. Since 1980, Benbow has been studying seventh-grade students who score in the top three percent on standardized tests of verbal ability and has seen no significant difference between boys and girls.

One reason the males can keep up with the females on the verbal section of the SAT seems to be that the test, like IQ tests, has been designed consciously to minimize sex differences. Nearly one-fourth of the verbal portion of the SAT is devoted to verbal analogies, questions that look something like:

music : loud :: ???
(a) art : thought-provoking   (b) food : spicy
(c) television : comedy   (d) cheeseburger : tasty

For some reason, males tend to do better on such verbal analogies than females, and if these were removed from the test, girls might well outpoint boys on the SAT's verbal section. Indeed, the analogies hardly test language skills at all – they demand that a person recognize the meaning of relatively common words, something that males and females do equally well, but past that, getting the right answer depends on abstract reasoning. To answer the above analogy, for instance, one must figure out that just as some *music* is *loud*, some *food* is *spicy*, and that *loud* and *spicy* are both sensory qualities, one for hearing and the other for taste. Thus (b) is correct.

On standardized tests that evaluate language skills more extensively, such as the English subtest of the ACT (American College Test) and the writing test that can be taken with the SAT, girls score better than boys. This superiority may soon lead to females getting higher scores on the verbal section of the SAT as well. The Educational Testing Service, which designs the SAT, has announced that the test will include a writing sample portion starting in 1994. It's not too reckless to predict that when this happens, girls' scores will improve relative to boys', unless of course the Educational Testing Service includes even more verbal analogies to keep everything balanced out.

So why isn't the math portion of the SAT balanced in the same way? The answer seems to be that there is nothing like verbal analogies that could be added to the math test which would favor girls enough to balance out the male advantage on the other questions.

Even if the verbal portion of the SAT were to drop verbal analogies and add a writing sample, however, scores on it still wouldn't reflect the true extent of the female superiority in language. The biggest and most important sex difference in verbal ability is found not among the college-bound boys and girls who take the SAT, but rather among those at the bottom of the scale – the children who have trouble with basic speaking and writing skills. Of the four to five percent of the general population who stutter, for instance, males outnumber females by three or four to one. The difference is just as striking among dyslexics – children of otherwise normal intelligence who have difficulty learning to read. There are about three times as many boys as girls with severe dyslexia, so a large percentage of the children who have real trouble with reading are boys. Unlike the male advantage in math, which is most pronounced at the upper end of the scale, the female superiority in language skills is clearest at the low end.

Researchers have suggested a variety of social explanations for why females develop better language skills than males – parents tend to talk more to daughters than sons, for example, or perhaps girls are simply putting their intellectual energies into verbal development because they feel mathematics is closed to them. Those hypotheses may well explain some of the difference in verbal abilities, but clearly not all of it. So far, no one has suggested a plausible environmental reason for why eighty percent of all stutterers or seventy-five percent of all severe dyslexics should be male. It's hard to think of a good explanation that doesn't involve some biological factor.

The sex differences in math and, to a lesser extent, language, get much of the attention from scientists both because they're important and because in some sense they're 'easy'. Many standardized tests exist to measure math and verbal skills, and millions of schoolchildren take these tests each year, so scientists have plenty of numbers to crunch in hopes of finding something meaningful. With sophisticated computer programs a researcher can look at the test scores from every angle, applying statistical analysis to determine what is significant and what isn't, and in the end he'll have an objective, concrete measure of how the sexes compare on some particular ability: 'Girls (boys) are smarter than boys (girls) in _____ (fill in the blank).'

But such simple numerical comparisons don't tell the whole story. Sex difference researchers are finding that even in situations where males and females perform equally well, they may solve a problem in different ways.

One of the best-studied examples is also one that hits particularly close to home for me.

My wife and I play the same little game every time we go on a vacation, or even take a quick trip across town. It goes something like this: 'Will you please stop looking at that map so we can get going,' she says. 'Just a minute,' I say, 'I'm just trying to make sure I know how to get there.' 'That's what you said ten minutes ago,' she grumbles. 'It would be faster just to get lost than to have you playing with that map all day.'

One summer on a visit to Sweden we were walking along the docks of Marstrand, quite possibly the most beautiful little sailing village in the world. To one side of us was the deep blue water of the harbor, gently jostling a fleet of the prettiest sailboats I'd ever seen. On the other side were rows of clean white houses with orange roofs, a manicured park, and a series of stone streets winding their way up a hill toward a fifteenth-century fortress we could see above the rooftops. Amy was eager to take in as much of the city as possible and she was urging me forward, but I could only walk so fast with the map unfolded in front of me, trying to trace out our route with my finger. 'OK,' I mumbled, 'the sun is in the east, so we must be heading north. This street is Hamngatan and we just passed Drottninggatan, so Kungsgatan should be coming up on the left. After that, the street is going to veer around to the west and we'll be coming to...'

That was the last straw. For the past week Amy had been relatively tolerant of my map habit, allowing me to happily plot our course from one town to the next with my three or four road maps. But now we were on an island – a very small island, she pointed out – and there was a castle looming over us, serving as a landmark. We couldn't get lost, so why did I have to keep looking at the map? I didn't have a good answer to that, and since I thought I'd heard in her voice the first, faint rumblings of an approaching storm, I reluctantly folded up the map and promised I wouldn't look at it again for at least fifteen minutes.

The rest of the visit was pleasant and uneventful. We climbed a hill to watch the sailboats heading out to open sea, toured the castle, shopped for T-shirts, and wandered up and down most of the streets on the island. The few times we weren't quite sure where we were, Amy could usually look around, spot a building that we had passed half an hour before, and point us in the right direction before I had even had time to open the map. Of course I would still unfold it and study it for a minute – just to make sure I know where we're going, I'd tell her.

I've always loved maps – I like to plot out my course before getting started and then periodically refer back to the map to check my progress – but before I began to study sex differences it never seemed more than just a personal foible. Now, after talking with Thomas Bever, I realize that I'm not alone.

Bever, a psychologist at the University of Rochester, has performed a series of experiments exploring how men and women learn their way around a new place. Since it would be too expensive to send his subjects to Sweden, provide them with maps and surreptitiously follow them around as they look for a good smorgasbord, Bever came up with the next best thing: a computer maze. Each subject sits in front of a computer and watches the screen, which displays a picture of a hallway receding into the distance and corridors branching off to either side. By pushing keys on the computer keyboard the subject can move down the hallway, turn right or left down the corridors, and even reverse direction. It's not quite the same as walking around a Swedish village, but Bever does provide 'landmarks' to help people find their way through the maze. Instead of parks, shops and seaside homes, the maze's landmarks are simply letters marking each intersection. And to serve as general points of reference – something like the castle on the hill in Marstrand – the maze also includes a stylized sun in the 'east' and moon to the 'west'.

In the experiments, many of them done with the help of his student, Dustin Gordon, Bever gives the subjects half a dozen practice runs to learn the route from start to finish, and then he times them. Males and females do about equally well in this part of the experiment, Bever says. Next he modifies the maze by either removing the landmarks or changing the maze's dimensions, and again he measures how fast the subjects make it through. Here he finds a striking disparity.

With the landmarks gone, the women have a much harder time finding their way around the maze, but the men don't seem to notice that they're gone. Taking away the sun and moon also bothers the women, although somewhat less, while it hardly fazes the men – their times are still nearly as good as before. But when Bever changes the geometry of the computer maze by lengthening some of its corridors, suddenly the men's times slow down and it is the women who are unaffected.

These results, Bever says, point to a basic difference in the way that males and females find their way around a place. Women tend to rely on landmarks to locate themselves in a new place, while men construct a mental map made up of 'vectors' indicating the distance and direction from one point to another. Thus removing the landmarks hurts the women's performance, while men are confused when the distances shift.

This and one other related experiment have led Bever to conclude that maps really are a 'male thing'. In that study, he had male and female subjects first learn their way around a physical maze, and then, when the men and women were equally adept at getting from the beginning of the maze to its end, he tested whether they had envisioned maps of the maze in order to navigate it. Giving the subjects a choice of eight maps, some of which didn't look anything like the real maze, he asked each to pick out the one map that represented the maze they had been exploring. Eighty-seven percent of the males got it right, but only twenty-five percent of the females did.

So now if Amy kids me about playing with maps, I just tell her, 'It's a male thing – you wouldn't understand.'

In place of maps, females tend to learn many more details about an area and, like Amy in Marstrand, can use those details to locate themselves. Bever discovered this when he switched his attention from computer mazes to the University of Rochester's underground tunnel system. Most of the university's major buildings are connected by tunnels which students use when the snow is too deep to walk outside. These tunnels have much in common with Bever's computer maze – the corridors are long and straight with an occasional intersection with another walkway, each corridor looks pretty much the same, and there are few obvious landmarks.

Bever reasoned that if women depend more on landmarks than men to find their way around, then the female students should be more sensitive to various features of the tunnels and be able to pick out 'landmarks' in the tunnels better than male students. So he took pictures at several dozen locations in the tunnels, showed them to students and asked them to pinpoint where in the tunnel system the pictures were taken. Although the men had trouble telling one spot in the tunnels from another, the women could identify many of the locations – especially the intersections, where they would have to pay attention to know which way to turn. 'I don't know how they do it,' Bever says. 'One place looks pretty much like another to me.'

Canadian researchers Liisa Galea and Doreen Kimura found a similar female superiority when they tested college students on a landmark-laden map. As a subject watched, one of the experimenters would trace out a complicated route on the map, which included compass directions, street names and drawings of various landmarks, such as houses, schools, shops and even a windmill. Then the researcher asked the subject to retrace the route from memory. Although the men made fewer mistakes in learning the routes, the women remembered more of the landmarks both along the route and off of it.

A big part of this sex difference seems to stem from what men and women think is important to remember about a place. Everyone, males and females alike, forms some sort of 'mental map' when learning an area, but the sexes include different details in these maps.

In 1983, Diane McGuinness and Janet Sparks tested eighteen male and eighteen female students at the University of California at Santa Cruz in order to learn which details they included on their mental pictures of the campus. They gave each subject a sheet of paper and asked them to draw maps, imagining that a friend was coming to visit the campus for the first time and would need a guide to get around the campus alone. Then the two researchers scored each map, counting the numbers of roads and sidewalks, buildings, and so on.

All of the students included major campus buildings such as the library and bookstore on their maps, but past that, the sexes performed very differently. The male students included nearly twice as many roads and paths as the females did. The female students made up for it by putting in many more extra items besides the major buildings – peripheral buildings, tennis courts, soccer fields and the like.

McGuinness and Sparks concluded that the men and women were using differing strategies to map out the campus in their heads: 'Females focus more on the landmarks ... and males focus more on the topographic network of roads and other connectors which provide a geometric framework for the location of buildings.' In other words, landmarks are more important to females and the broad outlines of a place are more important to males.

This difference really comes out when men and women give directions on how to get somewhere. In 1986, Leon Miller and Viana Santoni at the University of Illinois at Chicago tested direction-giving in a group of sixth-graders and a group of college students. Each group was shown a simple map with streets, various landmarks (houses, stores, restaurants, gas stations and so on), a scale bar indicating distance, and an arrow pointing north. After ten minutes of study, the map was hidden and the subjects were told to pretend they were giving their best friend directions: 'Your friend is at the Three Lakes train depot. What directions would you give him/her so that he/she can get to the Three Lakes college?'

In both groups the females mentioned about twice as many landmarks in their directions as the males, while the males gave twice as many distance estimates ('Go three blocks' or 'It's about half a mile from the gas station'). In a similar study done the same year, three psychologists at Temple University in Philadelphia found that males gave more mileage estimates and more cardinal directions (north, south, east, west) than females, although they didn't find a difference in mentioning landmarks.

So if you're planning to visit those married friends who live in another city, you might want to think about whether to ask the husband or wife for directions. Would you rather 'head north on Main Street for a mile and turn west on Elm' or 'drive down Main Street past the Burger King until you see the mall and then turn left'?

Before his work on computer mazes, Bever and his student Pietro Micceluchi discovered a very different sort of sex difference – one in how men and women learn. Bever had hypothesized that humans learn best when they're presented with the same information in two different ways. At first the brain struggles to resolve the two representations, but at some point they suddenly come together to form a single perception that is more complete than either view by itself. If he was right, it would point the way to more effective teaching methods.

To test his idea, Bever decided to teach subjects a new language. Some would only listen to it, others would only speak it, and a third group would both listen and speak. His theory predicted that the two-way group would learn faster than the one-way groups even if all three spent the same amount of time practicing.

Because a true foreign language such as German or Swahili would be too complicated to teach in a short experiment, Bever invented a simple artificial language based on English but with a different grammar. With only a few words and a few grammar rules to learn, the subjects could be taught the language and evaluated on it in a single session.

When he tested a large group of subjects, however, he discovered that the men and women performed very differently. The males did as he had predicted – they got about ten percent more correct answers after two-way training than after one-way. But the females were just the opposite – they did better when they practiced by listening or speaking alone.

That blew Bever's original hypothesis – it turned out to be true only for the male half of the population – but it uncovered a sex difference that no one had ever predicted. Men learn better if they get information in two complementary ways but women prefer consistency, at least in this limited type of language learning.

Bever found a similar sex difference in how rats learn mazes – the males learned better if they were taught the maze in both directions, but the females did best if they ran the maze in only one direction. This maze work was what eventually led Bever to the computer maze experiments on humans that uncovered the sex difference in landmarks versus compass directions.

In all of these examples, from computer mazes to mental maps to one-way versus two-way learning, the sex difference is not so much a matter of 'better or worse', but really just 'different'. One approach may give better results in some situations, the other may be superior in others.

Some researchers have suggested that sex differences in such things as mathematics may be at least partially caused by a difference in how men and women approach a task. In trying to solve word problems, for example, females may take a more verbal approach, trying to reason them out with words, whereas males may quickly resort to numbers. Since a numerical approach is usually more efficient for solving such problems, this could explain the male advantage on tests of this type.

McGuinness and Sparks describe another example of the different ways that men and women approach problems. They spoke with an international town planner who, during his twenty years of teaching students how to plan cities, had discovered that men and women approach laying out a city in very different ways: 'Females typically began their plans by delineating areas for specific purposes (residential, factory, school, etc.) and later connecting the buildings with roads, or omitting them altogether. Males more often did the reverse and began by carving up the site with a grid of roads – arranging the buildings as a secondary consequence of the road system.' It turns out that neither the male nor the female approach to laying out a town is very effective and that a good town planner has to find a strategy somewhere between the two extremes. There's probably a good lesson here somewhere.

Understanding these sex differences may eventually lead to new teaching techniques, some researchers suggest. Bever's discovery of the one-way/two-way difference, for instance, implies that it could be effective to use different methods tailored to either males and females in teaching such subjects as languages or mathematics. Some students –

mostly males – would do better if they had material presented to them in two different ways, while other students – mostly females – would be best taught from a single perspective. And recognizing women's preference for landmarks could point to a new way of designing maps. Including drawings or pictures of prominent landmarks might make maps more practical, for men as well as women.

Females excel at language, males have an advantage in math, and the two sexes sometimes use different approaches to solve the same problem . . . but how does it all add up? Which sex is really better, faster, smarter? Inquiring minds want to know.

Already we have seen that men and women differ in certain characteristic ways, and the next chapter will detail many more differences, so it's natural to compare the two, to try to keep score. OK, you might say to yourself if you're male, maybe women *can* write a better paragraph, but men are faster at complicated math problems. Or if you're female: sure, men can look at a map and know where they're going a bit more quickly, but women remember better where they've been.

That's the wrong approach. One of the clearest lessons from sex difference research is that it's silly to treat all this as some sort of contest. For one thing, the sex differences are statistical in nature and say nothing about an individual male or female. There are women who shine at math and men whose skills lie in language, girls who like maps and boys who don't. Indeed, many of the differences described in the coming pages are small enough that they're not even detectable unless you test dozens or hundreds of people at once. If you pick five male and five female high school students at random and ask each of them to write a short essay on saving the planet, chances are you won't notice any difference in quality between the boys and the girls. It's only when you do the same thing with a hundred students of each sex and grade them on a uniform basis that you start to find the girls holding a slight, but consistent advantage.

As I've been writing this book, my wife has made a habit of pointing out all the differences between us that don't fit with the stereotypes or the averages. 'Are you going to put *that* in the book?' she asks. Once we were driving home from a distant shopping center at night and I found myself completely turned around, but her sense of direction was perfect. 'Are you going to put *that* in?' And when we were talking about our childhoods, she discovered that she had spent a lot more time playing baseball, football and other team sports than I did. 'Are you going to put *that* in?'

Believe me, it's not hard to find plenty of exceptions to the general rules about male and female behavior.

An even more important reason not to keep score between the sexes is that it's impossible to place an objective value on such different traits as, for example, verbal ability and mathematical/scientific skills. Who's to say whether Hemingway or Madame Curie made a greater contribution to humanity? (Or for that matter, how can you compare the contributions of someone who stays home to raise a family versus someone who goes to work in an office?) And unfortunately, whenever someone does try to rank different human skills as more or less valuable, it always seems that the 'masculine' attributes come out on top. The statistics about boys doing better than girls in higher math are usually seen as threatening – as if they prove that females aren't quite as smart as males. But no one ever concludes that females must be the smarter sex because boys have more trouble learning to read and never quite catch up with girls on verbal composition, even though reading and writing are much more important to most people than calculus or geometry.

Talking with my wife about this book has really brought home to me just how ingrained this attitude can be. We're both bright people, but our patterns of intelligence are quite different. Much of my talent lies in math and science, while Amy is more verbally oriented. She's much better spoken and quick on her feet than I am, and when we're in conversations with other people I feel like a Model T in the Indianapolis 500. She reels off a series of insights or comments, each so carefully crafted that it sounds like she's reading from a script, while I'm struggling to focus one idea well enough to speak it aloud.

And she has another ability that leaves me completely awed. Psychologists call it 'behavioral intelligence' and, roughly, it's the ability to understand what other people are thinking or feeling and to respond appropriately. Researchers have paid little attention to it, but a few experiments have found a female advantage. Women are much better than men at looking at pictures of people and being able to read the emotions in their faces, for instance, as well as being slightly better at recognizing faces of people they know. And I can testify that in our household all of the behavioral intelligence is concentrated in one head, and it's not mine.

Amy is working on her Ph.D. in Public Policy, and one night shortly after the beginning of school she came home and told me what she had figured out about some of the other students. She hadn't met or talked to these people, only watched them in class. 'This one guy is wondering whether he really belongs in the program,' she said. 'Whenever he answers a question he tries to sound very intellectual, but he's trying to convince himself more than the rest of the class. There's a man and woman who sit next to each other – they're trying to decide whether to have an affair.' And so on. From past experience, I know she's probably right about all of them, and that amazes me. I could be sitting right next to somebody who had just won the lottery and I probably wouldn't even notice he looked a little excited.

Try this question on your friends: 'If you're at a party with a group of married friends and two of them are having a secret affair, would you notice?' I've found that women usually say yes and men say no. And of course, the best matchmakers are always women. At a wedding earlier this year, Amy and I met the woman who had brought the bride and groom together. Among the guests at the wedding was another couple, scheduled to be married a few months later, whom this woman had also introduced, and we heard that she could claim credit for two other weddings as well. It was just a hobby for her, figuring out which of her many female friends would fit well with which of her many male friends and then introducing them.

The point, however, is not so much that women have more behavioral intelligence than men – that's probably true although there hasn't been enough study to say for sure – but rather that this 'feminine' skill doesn't get much respect. When I told Amy she deserved a Ph.D. in behavioral intelligence, her response was, 'Great. I'd rather be able to do math.' Nor does she seem to value her own verbal talents as highly as the quantitative skills of other (mostly male) students in her graduate program.

She is, I'm afraid, reflecting the opinion of society in general, and this attitude is one of the great stumbling blocks to accepting the existence of sex differences. Too often, the ways in which women are found to be *different* from men are interpreted as ways they are *inferior* to men, as if the male is the standard by which all humanity should be judged. But the lesson that has gradually been taking shape over the past decade is that different doesn't imply unequal.

Sociologist Alice Rossi makes an important distinction about the idea of equality with regard to the sexes. Over the past few decades, quite a few people have confused 'equality' with 'identity' and assumed that if women are to be considered equal to men they will have to be seen as identical to men – the same innate mathematical ability, comparable aggressiveness, the same way of interacting with other people, similar goals and values, and so on. However, as Rossi writes, 'there is no rule of nature or of social organization that says men and women have to be the same or do the same things in order to be socially, economically, and politically equal.' In the past the problem has been that women were not socially, economically or politically equal, and various presumed sex differences were often used as a rationalization – women just aren't as smart as men, or not as competent, or not as ambitious, or whatever. And the understandable reaction to this state of affairs was to insist that women are not different from men at all and that if they were treated the same as men and given the same chances, all the sex differences (except the physical ones, such as who gives birth) would fade away.

Understandable as it may have been, the reaction was wrong. Scientifically wrong because the more we learn, the more obvious it becomes that there are innate differences between the sexes. And socially wrong because the practical effect of insisting on sexual identity was not to assert that 'men and women are the same' but rather to claim that 'women are the same as men' – a very different proposition which measures females against males but not vice versa. That is why you hear a lot about girls' mathematical skills in high school but little about boys' reading skills in grade school. And that is why progress toward sexual equality is measured by the number of women becoming doctors or politicians, not by the number of men becoming nurses and teachers.

It is not necessary to deny the existence of sex differences in order to insist that the sexes are inherently equal and that they should be treated that way. Indeed, the picture that emerges from sex difference research is of man and woman as two different but equal manifestations of humanity. Yes, women are better at certain specific skills than men, and men are better at certain specific skills than women, but there seems to be a rough equality in the way these differences are apportioned. Neither sex has a monopoly on desirable traits. And furthermore, most of the sex differences have no clear winner or loser. Is it better to use landmarks or a directional sense to find your way around? Is it better to be aggressive or nurturing? Who's to say?

Sex difference researchers agree that none of the dissimilarities between the sexes should be seen as threatening to anyone, male or female. They may demonstrate that men and women are not the same, but never that they are not equal.

CHAPTER 2

A Tale of Two Sexes

Sitting in the living room of her home on Sanibel Island in Florida, Diane McGuinness is trying to explain why she became interested in sex differences. It probably had something to do with her ex-husband, she jokes. 'I had grown up in an all-female household,' she says, and when she got married, 'I found myself living with this alien thing. I couldn't understand why he did what he did.' But no, she decides, it actually goes back much further than that.

McGuinness started playing the piano when she was eight, and by the time she was twelve she was giving piano lessons herself, a teaching career that continued for nearly thirty years. And over those years, she says, she saw a big difference in how boys and girls approached learning the piano. 'When you start a little boy off by showing him where middle C is, he'll play it once or twice, then he'll get up and start looking all over the piano.' Now McGuinness herself has jumped up and run over to the piano, where she proceeds to act out the part of the little boy. 'He'll climb up on the bench and reach inside the piano,' she says with her own head stuck underneath the piano's lid, 'and he'll pluck the strings and push on the hammers.' Plink, plink, plunk go the strings. 'Little girls never did that,' she says as she pulls her head out and sits back down.

'And if you praise a boy for playing something well, he'll do something like this,' she says, raising her fist in the air and letting out an exultant 'Yesss!' This was true even thirty years ago, she adds, before athletes in televised events made such a big production of congratulating themselves on a good play. And how do girls react to praise? Still acting out her answers, McGuinness puts on a shy little girl's smile, holds her arms close to her body and shrinks into herself as if embarrassed by the attention.

McGuinness, a professor of psychology at the University of South

Florida in Fort Myers, was a musician before she became a scientist. Her first degree was in English and music, and for nearly a decade she taught music and performed as a concert singer in England. But in the late 1960s she went back to school at the University of London's Birkbeck College to get a bachelor's degree in psychology, and she continued on to get her Ph.D. from University College at the University of London in 1974. Why the career change? 'I had always seen different things in people and wondered why they were like that,' she says, but she didn't become seriously interested in psychology until she saw the tremendous changes in her own mother's personality caused by severe hypoglycemia.

For a period of several years, McGuinness says, her mother had an undiagnosed disease that caused abnormally low levels of sugar in her blood, and McGuinness had watched her go through a 'radical personality change' that included delusions of people plotting against her. Once the hypoglycemia was recognized and controlled, her personality bounced back to what it had been. McGuinness was intrigued by the fact that something as simple as the level of glucose in the brain could change a person so dramatically. 'That happened sometime around the time I was trying to decide whether to continue pursuing my musical career,' she says, and one day she woke up to find she had made the decision to go into psychology.

Back in school as an undergraduate, McGuinness did a senior thesis in psychophysiology, a branch of psychology that studies how the body's physical status is affected by various states of mind, such as how a person's heart rate and respiration change when the person is frightened. Entering graduate school, however, McGuinness decided she wanted a broader sphere to work in. She wished to 'integrate and synthesize' information from a number of areas, and sex differences, which had been the focus of her senior thesis, looked like a broad enough subject to hold her attention for many years. 'If I concentrated on that, it would allow me to be a dilettante and wander all over the landscape.' Now, two decades later, she has done just that – her list of publications includes an array of imaginative experiments on a broad variety of subjects, from sex differences in hearing and vision to how men and women draw maps or interact in groups.

Anyone who has been around small children has discovered, as McGuinness did, that boys and girls are different. Every parent can add his or her own anecdotes to the litany – little Susie was talking up a storm when she was fourteen months old but Sam hardly said a word until he was three, or Jason is always running around, throwing things and fighting with his little brother while Jennifer is content to sit in her room and read or play with her dolls. My own nieces and nephews are just like that. I remember one visit where four-year-old Abbie was content to cuddle in my wife's lap for almost an hour while Richard, six, and Timothy, two, were in the other room crashing trucks together, seemingly just to see how much noise they could make.

And it's not just kids. Like McGuinness, most adults have at some point wondered about some 'alien thing' masquerading as a member of the opposite sex.

But exactly what are the differences between the sexes? They are much more extensive than simply a male advantage in math and a female superiority at verbal skills, or different ways of findings one's way around a strange place. They are not, however, always what the stereotypes would have you believe.

In the early years of this century, for example, it was commonly believed not only that athletic competition was unfeminine but that certain events were actually harmful to women. You can track the evolution of attitudes toward women's sports by the women's events included at the Olympic Games. There were none in the initial Games of 1896, then golf and tennis in 1900, archery in 1904 and figure skating in 1908. By 1912 the Olympic Committee had decided women were up to the rigors of competitive swimming (although no more than a hundred meters), but it wasn't until 1928 that track and field events for women were okayed. Even so, the 800-meter run was canceled after that year and not reinstated until 1960 because it might be dangerous for women. There was no women's marathon at the Olympics until 1984. Today, when the top women marathoners finish the twenty-six mile race faster than all but a handful of men and female track athletes run races faster than the best men did just a few decades ago, it seems silly to have thought women any less capable than men at this type of activity, but such ideas were taken seriously not too long ago.

The moral of the story: sometimes what we 'know' about the differences between the sexes turns out not to be true at all, and is often no more than a self-fulfilling prophecy. If women aren't encouraged or allowed to run, jump and throw, it's hard for them to prove that they can.

Fortunately psychologists have recognized this problem, and over the past twenty-five to thirty years many of them have worked on pinning down the differences between the sexes, independent of myths and stereotypes. Some of the findings do agree with the old stereotypes, but many don't, and the emerging pattern of differences offers a new and truer picture of man and woman, boy and girl.

The most obvious differences are the physical ones, and those are well known. Men are, on average, about nine percent taller than women – in the United States the average man in his twenties is nearly five-foot-ten and the average woman just over five-foot-four. Thanks to testosterone, males are more muscular than females from puberty on, with muscles accounting for an average of about forty percent of a young man's total body weight but only about twenty-three percent of a young woman's. Testosterone also triggers an adolescent boy's deepening voice and growth of facial and body hair. In females, estrogen at puberty prompts the development of breasts and a widening of the hips. The wider pelvis will cause a woman to sway her hips more than a man when she walks. And the two sexes tend to collect fat in different places – mostly around the waist and stomach in men but more on the hips and upper legs in women.

Surprisingly, there are also consistent sex differences in all five senses: sight, hearing, touch, taste and smell. One of McGuinness's first studies, for instance, looked at differences in how men and women respond to sound. At the time it was already well known that females hear high-pitched sounds better than males, with the difference beginning in childhood and getting larger as people get older. McGuinness looked at another issue: how sensitive males and females are to loud sounds. She asked twenty-five males and twenty-five females to adjust the volume of a tone until it was just a little too loud to be comfortable. On average, the males turned the volume up eight decibels louder than the females, and this was true at every frequency from deep bass to high treble. To put this into perspective, McGuinness notes that increasing the volume of something by approximately nine or ten decibels makes it seem twice as loud, so males are comfortable with sounds nearly twice as loud as females are. This may explain, McGuinness says in all seriousness, why a man may turn up the volume of a televised football game so loud that it hurts his wife's ears.

Could this difference be caused by boys being 'taught' to put up with louder sounds? Perhaps boys are allowed to make more noise when growing up so they get used to louder sounds, or else males think it's 'macho' to put up with a little pain in the ears, but McGuinness doesn't think so. She notes that another researcher found just as large a sex difference in volume tolerance among school children at five to six years old and at ten to eleven years old. If the tolerance developed over time, the younger children should show a smaller sex difference. The difference seems to be innate, not learned, McGuinness says, and this greater female sensitivity to sounds conceivably could play some role in the female advantage in language skills.

For sight the sex difference is reversed, with males more sensitive to bright lights than females. Under normal lighting conditions, men have slightly better 'visual acuity' on average than women – they can see smaller details, such as the fine print on an optician's eye chart. And this difference in acuity becomes bigger when the target is moving. In one study of seventeen thousand people getting driver's licenses in California, men and women were shown a small, rapidly moving square and asked to say which corner of the square held a checkered pattern. With each trial, the square got smaller and smaller until the subject could no longer pick out the pattern. On average, men could pick out finer details on the moving square than women, and the faster the square moved, the larger the male advantage became.

The same researcher, Albert Burg, also studied the driving records of his subjects and discovered that, on average, people with better 'dynamic visual acuity' had fewer accidents – which might help explain why men have fewer accidents per mile driven than women. Burg found that although the male drivers in his sample were more reckless, to judge from a greater number of traffic tickets per mile driven, they still had a lower accident rate than the female drivers. This agrees with nationwide statistics which find that women are involved in twelve percent more accidents per mile driven. That doesn't mean, however, that you should hand over the car keys to the male in the family (especially if he's a male teenager, for whom the accident rates are practically off the chart). Although men are slightly less likely to be involved in the minor fender benders, they are sixty percent more likely to be involved in fatal accidents.

In the dark, the sex difference in vision is reversed. McGuinness found that in a completely dark room women are much better than men at picking out a faint patch of light. Females' eyes adjust to the dark faster than males' eyes, and they also hold the after-image of a briefly lit object longer, which is an extra advantage in a very dark setting. In yet another study on vision, McGuinness showed that men have sharper vision in the middle part of the visual field, but women have better peripheral vision.

This might, I suppose, account for the observation that so many children have made while growing up – that mothers seem to have eyes in the backs of the heads.

Men are more likely than women to be color-blind because the gene for color vision is carried on the X chromosome and females have two X chromosomes while males have only one. If a male's single gene for color vision is defective he'll be color-blind, but a female has a back-up gene and both genes must be out of commission for her to be affected. Otherwise there appear to be no major differences in color vision between the sexes, McGuinness says, although women may be slightly more sensitive to red colors.

Touch, the most intimate of the senses, shows probably the most consistent sex difference. In one study a researcher measured sensitivity to touch by using thin filaments that were pressed down on the skin very lightly at first and then with increasing force until the subject could detect the pressure. Measured this way on twenty spots from head to toe, females were more sensitive than males at every point except the tip of the nose, and the female advantage was particularly large on the arms, legs and trunk.

Women also have a slight but consistent edge in smell, says odor researcher Richard Doty at the University of Pennsylvania. Not only can women detect fainter odors than men, but they're better at recognizing the smells that they detect. Doty has developed a scratch-and-sniff, multiple-choice test that offers a subject forty odors to identify. For each odor the subject is given four choices that smell nothing alike – pizza, bubble gum, rose and natural gas, for example – and then asked to pick out the correct smell. Among men and women in their twenties, both of whom are very accurate on this test, females are slightly less likely to make mistakes. By the time people reach sixty-five, the male performance has dropped sharply but women are only slightly worse than they were at twenty-five.

It's the same story for taste, Doty says. Women can detect lower concentrations of sweet, sour, bitter and salty tastes, although the difference is not great.

But of course the more interesting question is, Do men and women naturally like different foods? The answer is yes, but the sex difference in taste preference is not always what you'd expect. Despite all the stereotypes about women and chocolate, for instance, it's actually males who go for the sweet in taste tests. One study asked males and females to evaluate the taste of soft white cheese or heavy cream that had been sweetened with sucrose. The females gave the highest taste rating to samples with a sucrose level of about ten percent – anything more was too sweet – but males preferred food with about twice as much sucrose. And the younger the male, the sweeter he liked it: small boys didn't think that even forty percent sucrose was too sweet. (At our house, I can't understand why my wife likes diet sodas, and she thinks it's sickening when I have a big slice of pecan pie and wash it down with a Coca-Cola.)

Adam Drewnowski, a nutrition expert at the University of Michigan, reports that both men and women like fatty foods, but they prefer their fat in different forms. Females choose foods high in both fats and carbohydrates – doughnuts, cookies and cake – while males are more likely to pick foods that have plenty of fat and plenty of protein. This means steaks, roasts, hot dogs, hamburgers and lunch meats.

Apparently because more men than women are what psychologists call 'sensation seekers', always out for a thrill, men are much more likely than women to enjoy hot, spicy foods. And the same reason seems to explain why men are more likely to seek out new and unusual foods – things like pig's feet, raw eggs or sardines.

Besides the five senses, men and women also differ in muscle control. Women have better manual dexterity and more precise control of their arms and hands than men. That gives them an advantage in playing a musical instrument, typing and assembling small objects, and also in the delicate cutting and stitching necessary for either fine embroidery or brain surgery.

Males, on the other hand, have faster and better controlled 'big' movements such as running and jumping, throwing and catching. You might guess that this advantage arises from boys spending more time playing sports and physical games, but the research doesn't bear that out, says Doreen Kimura, a psychologist at the University of Western Ontario in London, Ontario.

In 1990 Kimura and a colleague, Neil Watson, tested the throwing and 'intercepting' skills of four dozen university students, evenly divided between men and women. In the throwing test, the subjects threw a small dart at a target some ten feet away, with twenty trials for each hand. In the interception test, a ping-pong ball was fired toward the subject from one of a dozen launching devices about fifteen feet away. The subject tried to stick out a hand to intercept it as it went by, and was given thirty chances with each hand. The men did much better than the women on both tests and for both hands, with the male advantage being slightly larger for the right hand and for interception.

This advantage, Kimura and Watson said, seems likely to be innate and not caused by boys having had more practice in sports, for several reasons. First, throwing small darts and intercepting ping-pong balls are very different skills from those developed in baseball or football, and the longer arms that give males an advantage in throwing a baseball farther and faster might actually make it harder to hit a close target since longer arms would increase the chance for error. And even if throwing a baseball did make boys better than girls at throwing darts with their right hands, it shouldn't have made a difference to the left. (All of the subjects were right-handed.) Furthermore, Kimura and Watson gave a sports history questionnaire to all the subjects asking how much experience they had playing various sports, and the two researchers found almost no correlation between athletic experience and how well a subject did on the tests. And finally, in an earlier study, Kimura had tested preschoolers and grade school children and found that boys are consistently more accurate throwers from an early age. If constant practice by boys were the reason for their superiority in throwing everything from baseballs to darts, then that edge should be small or zero in very young children, Kimura says, but that's not the case.

In terms of physical health, women are superior to men on almost every count. Everyone knows that women live longer – a female born in the United States can expect to live eight years longer than a male born at the same time in the same hospital – but not everyone knows that the female advantage starts at conception. There are probably about one hundred and twenty males conceived for every one hundred females (estimates range from one hundred and ten to one hundred and seventy), but since a male fetus is much more likely to undergo a spontaneous abortion, the ratio at birth is only one hundred and five males to every one hundred females. No one knows why more male fetuses die, but a likely explanation is that the male fetus must go through an extra stage of development – the masculinization caused by testosterone – which offers one more chance for things to go wrong.

After birth, boys are more likely than girls to die from disease and accidents, and by puberty the ratio of males to females is about even. The trend continues throughout adult life, with the result that the over-eighty population is predominantly female.

Scientists don't know exactly why men die so young, relative to women, but it seems likely to be a combination of biological, environmental and social factors. Biologically speaking, the fact that women have two X chromosomes and men have only one means that women are less sensitive to genetic defects on an X chromosome, and some researchers have suggested that males may have a higher metabolic rate which causes their bodies to wear out sooner than females' bodies. On the social/environmental side, more men than women die from such things as lung cancer (thanks to more men smoking, and smoking more), auto accidents, suicide (more females attempt suicide, but males are much more likely to be successful), cirrhosis of the liver (men drink more) and heart disease (apparently a reflection of sex differences in diet and stress). If women were to smoke and drink more, drive more recklessly, take riskier jobs (coal-mining, fire-fighting and so on), get more stressed out about their work and internalize that stress, and then insult their bodies with the same high-cholesterol foods that men do, they might start dying at the same rates that men do.

You might predict the same thing would be true for ulcers, which strike men much more often than women – perhaps as women win more of the high-paying, high-stress jobs they'll start getting ulcers as often as men. But then again, it's possible that males are simply more prone to developing ulcers. At least that's one possible implication of a study on mice done a few years ago by a group of researchers in Japan. The researchers found they could cause ulcers in the rodents by putting them into a box and giving them a short electrical shock every two minutes for twelve hours. Surprisingly, a second group of mice that weren't shocked but simply watched and listened to the reactions of the first group developed almost as many ulcers as the ones who got the electric shock. When the researchers analyzed the ulcers by sex, they saw that almost all of the male mice in both the physically shocked and the psychologically shocked groups had developed ulcers, but less than a third of the female mice in the two groups had, and the ulcers were more severe in the males.

To see what might be causing the sex difference, the researchers castrated some male mice just after weaning. Granted, castration sounds like it might cause ulcers all by itself, but when these mice eunuchs were subjected to the same shock treatment as the others, they actually developed many fewer ulcers than intact male mice and about the same number as females. It seems, the researchers concluded, that testosterone produced by the testes somehow causes mice to develop more ulcers when they are stressed. The scientists stopped short, however, of suggesting the obvious treatment for stressed human males who want to ward off ulcers.

None of the physical sex differences provokes much controversy since it's impossible to deny their existence or explain them away by differences in how males and females are raised, and anyway, they have few implications for social policy. But it's a different story for mental and psychological sex differences. These do have implications for how a society should be structured – Are there jobs for which one sex is likely to be more qualified than the other? Are women better at raising children than men? – and at the same time they cut right to the core of a person's identity and sense of self-worth. Most women don't mind that males can lift heavier weights than females, but few are comfortable with the suggestion that men may be naturally more competent in certain areas of mathematics. Men don't worry that they can't breast-feed a son or daughter, but many new fathers have wondered just how nurturing they can be. Concerns like these make the study of mental and psychological sex differences exceptionally sensitive.

At the same time, it's much harder to pin down exactly what these differences are since they're more difficult to measure objectively than physical differences. A tape measure works just fine for comparing heights, but how do you test which sex is more concerned with helping others?

Complicating things even further is how society creates or exaggerates sex differences. If Steve is six inches taller than Cecilia, it's not because he got more encouragement to grow, but if he gets straight As in calculus while she's uncomfortable with percentages, how much of that is innate and how much is due to Cecilia being told that girls can't do math? Of all questions about sex differences, the debate about what role society plays and what role biology plays is the most difficult to answer, and much of the rest of this book will center on that. For now, however, we'll be content to ask merely, What are the differences?

We can start out by saying what some of them aren't. Nearly twenty years ago, in their landmark book *The Psychology of Sex Differences*, Eleanor Maccoby and Carol Jacklin began the summary chapter with a list of widely held beliefs about sex differences that they had found no evidence for. That list bears repeating since many of those beliefs are still widely held today among the public, if not among scientists.

Myth 1: 'That girls are more "social" than boys.' Maccoby and Jacklin found that males and females are equally interested in being around other people and that in childhood girls do not spend more time – and sometimes spend less time – with playmates than boys do.

Myth 2: 'That girls are more "suggestible" than boys.' The sexes are 'equally susceptible to persuasive communications and in face-to-face social-influence situations,' Maccoby and Jacklin concluded.

Myth 3: 'That girls have lower self-esteem.' Females report just as much self-satisfaction and self-confidence as males through adolescence, and the few studies of self-esteem later in life found no sex differences. In college, males do have a greater sense of being in control of their own destiny than females and also express greater confidence in how well they will do in school, but even then their female peers have no less self-esteem than they do.

Myth 4: 'That girls are better at rote learning and simple repetitive tasks, boys at tasks that require higher-level cognitive processing and the inhibition of previously learned responses.' No consistent sex difference was found in a variety of psychological tests designed to measure this.

After dismissing these and several other myths, Maccoby and Jacklin asked why they should be so persistent when there is no scientific evidence for them. It's possible, they wrote, that scientists just haven't figured out the right way to detect sex differences that are obvious to normal people in everyday life, but if that's the case, the sex differences must be rather limited and not the sweeping dissimilarities of popular belief. Instead, they concluded, the real reason that people continue to believe these myths about men and women is 'the fact that stereotypes are such powerful things'. Once stereotypes get started about a group of people, they tend to be self-perpetuating: '[W]henever a member of that group behaves in the expected way the observer notes it and his belief is confirmed and strengthened; when a member of the group behaves in a way that is not consistent with the observer's expectations, the instance is likely to pass unnoticed, and the observer's generalized belief is protected from disconfirmation.'

In other words, if two women in an office stop to talk in the hall they're just being chatty, but if two men do it they're probably working on an important deal or, at the very least, they're 'networking'. If a woman loses her temper or cries, it just proves how emotional females are, but no one notices the woman who keeps her head in a crisis or who makes an important decision carefully and rationally. When a boy in school does well in math and science, it's just his native masculine knack for those subjects, but the boys who never made it past long division just weren't trying.

Still it's also true – as Maccoby and Jacklin discovered – that quite a few of the stereotypes about the sexes have some validity. For instance, psychologists have confirmed many of the things that parents have always known about children.

Boys are indeed more active than girls. It's already apparent in the womb and during the first year after birth, but the difference increases over time and reaches a peak in elementary and junior high school. The size of the difference depends on the setting (it's more pronounced when children play in groups, for instance) and there is plenty of overlap between the sexes – some girls are more active than most boys and some boys are less active than most girls – but on the whole boys move around more and expend more physical energy than girls.

That difference in activity is compounded by what Diane McGuinness calls different 'styles of play' among boys and girls. Most people know that boys and girls have their own toy preferences and that it doesn't do too much good for Mom and Dad to try to get their daughters to play with trucks or their sons to play with dolls. (My niece got a truck for Christmas last year, and when she opened the package she said, 'Daddy, you know I don't like trucks.' Her two brothers were only too happy to take it off her hands.) However, it's not the toys themselves that are responsible for the different preferences, McGuinness says, but what children can do with the toys. Boys like things they can manipulate and move around – building blocks, erector sets, cars, trucks, airplanes and so on – while girls get no special satisfaction from such movement and stick with dolls and arts and crafts. The female preference for dolls may reflect a natural inclination to play with and take care of infants (in many species of monkeys and apes the young females will play with infants while the young males ignore them), but this hypothesis remains controversial.

In 1990 McGuinness and several students watched thirty-eight boys and thirty-five girls, ages three to five, as they played without supervision in their preschool. Each child was observed individually for twenty minutes and his or her activities were carefully recorded. When McGuinness analyzed the data, she found striking differences in how the children spent their time.

On average, girls undertook only half as many activities as boys during the twenty-minute period and they spent much more time on each endeavor. Girls also worked more steadily on things, while boys were more likely to interrupt what they were doing and come back to it later.

Girls took part in more teacher-organized activities. Boys more often used a toy or other object for something other than what it was intended. More girls than boys chose painting as an activity, and girls spent an average of three times as much time on a painting as boys. Boys spent more time watching other children and were more likely to hit or push another child.

> Apart from these findings, there was a qualitative difference between the boys' and girls' behavior that is difficult to capture. More boys than girls appeared to wander aimlessly from place to place, from activity to activity, or child to child, as if in a state of bewilderment or confusion... Almost without exception, every boy interrupted what he had settled to do to investigate something else going on in the room. Often they were distracted by a noise, a child laughing, movement of children or a newcomer entering the room. On the other hand, there were times when we could discern no particular distracting event and no basis for the interruption could be found. One boy, for example, frequently interrupted play with a toy dump truck, which he returned to over and over again, by walking across to another child to say 'hello', to a different child to help with a puzzle, and for a sudden dash in and out of the play house.
> Another aspect of boys' behavior that resulted in a frequent shift from one activity to another was lack of closure. Four boys who asked to paint and were provided with fresh paper, water, etc. made a few cursory brush strokes and announced: 'All done!' A similar lack of completion was observed at the crafts table and a range of other activities which had obvious end points, such as card or board games.
> The observations on the girls indicated that once they settled to a task, most rarely put it down or quit until there was some level of completion. If they interrupted what they were doing, it was more often for a purpose: to show the teacher or another child, or because of some unusual event.

As they get older, children's styles of play change radically, but one thing remains the same: boys and girls play in very different ways. In the 1970s Janet Lever at Northwestern University in Evanston, Illinois, spent an entire year observing and interviewing one hundred and eighty-one fifth-graders from three schools in order to study sex differences in the games that children of this age play. Although the work is now two decades old it remains probably the most thorough look at the types of sports, games and recreation that boys and girls choose. Lever watched the children during recess, gym class and after school, gave questionnaires to them asking how they spent their time away from school, and interviewed a third of them one-on-one. She also constructed 'diaries' of a week's worth of after-school activities for each child by having them fill out each morning a list of what they had done the day before, whom they did it with, where it was and how long they spent doing it.

Lever documented six major differences in the ways boys and girls play. First, boys played outdoors much more often than girls, perhaps because the team sports and games like 'War' that boys like are best done outside while girls' favorite pastimes, dolls and board games, are inside activities. Second, boys played in larger groups than girls, again related to the types of activities they preferred. Third, groups of boys were more likely to contain participants from different age groups. Boys accept the idea that 'you're better off with a little kid in the outfield than no one at all,' Lever wrote.

> For example, I witnessed numerous ice hockey games where ages ranged from nine to fifteen or sixteen. The youngest children tried their best to keep up with the older ones or dropped out. They've learned to accept their bruises, stifle their frustrations, or not be invited to play gain. The very few times I observed girls in age-mixed play was at summer camp when the ten-to-twelve year olds used much younger children of five and six as 'live dolls', leading them in circle songs like 'ring around the rosy' or versions of tag like 'duck, duck, goose'. Here the oldest girls had to play on the level of the youngest child instead of vice versa.

Fourth, girls were more likely to play with a group of boys than the other way around. The boys accepted those girls who were athletically skilled enough to hold their own. By contrast, when a boy entered a girls' game like hopscotch or jump rope it was always 'in the role of "buffoon" or "tease" – there to interrupt and annoy the girls and not to be taken as serious participants.' Fifth, boys' games were more competitive – they had fixed sets of rules, a predetermined end point, and winners and losers. Sixty-five percent of the boys' after-school activities were competitive games compared to thirty-five percent of the girls', and the difference remained even when team sports were excluded. And sixth, the boys' activities lasted much longer. At recess, Lever found, '[b]oys' games lasting the entire period of twenty-five minutes were common, but in a whole year in the field, I did not observe a single girls' activity that lasted longer than fifteen minutes.' Apparently fifth-grade boys have changed quite a lot from the time they were four or five, when they seldom stayed with any activity for very long.

In a later study of children's games, Lever focused on another difference: boys' games are much more *complex* than girls' games. This complexity was apparent in a variety of ways. The boys' games generally had more rules than girls' games – baseball versus jump rope, for instance – and the boys were correspondingly more attuned to playing by the rules.

'A gym teacher introduced a game called 'newcombe', a simplified variation of volleyball, in which the principal rule is that the ball must be passed three times before being returned to the other side of the net,' Lever wrote. 'Although the game was new to all, the boys did not once forget the 'three-pass' rule; the girls forgot it on over half the volleys.'

The boys were more likely to play in games where participants took on different roles – pitcher, batter, catcher and so on – while girls more often chose 'single-role play' such as riding bikes or ice-skating, where all the players do the same thing. Boys' activities had more of what Lever called 'player interdependence' – one person's actions affect the next player. There is little or no interdependence between children riding bikes or taking turns at hopscotch, more interdependence in competitive games like checkers or tennis, and most interdependence in team sports where one must react to what both opponents and team-mates do. And boys preferred to include explicit goals – scoring the most points in a hockey game, for instance – so that a winner could be determined. Girls, Lever wrote, seem to be much less interested in who won.

Since children's play develops skills that a child uses later in life, Lever speculated that the sex differences in play may lead to boys and girls developing different social skills. The implications are obvious: boys learn to compete with one another, to lead and to follow, to work together with others on a team toward a common goal, often in competition with other teams, and to 'play by the rules', while girls learn to interact one-on-one or in small groups, to work co-operatively and to have consideration and empathy for others. The social skills taught by boys' games are very much like those needed for success in running business or working one's way up the corporate ladder, Lever noted, and while admitting that she had no proof of a connection between early play and later success, she noted that one study of women in top management positions in American companies found that most of them were former tomboys.

That situation may be changing, however. Twenty years ago, when Lever did her studies, a woman who succeeded in the corporate world probably needed 'masculine' business skills because she was working mostly with men and playing by their rules. Recently, though, there seems to have been a trend among businesses to encourage more 'feminine' skills in their managers and employees, such as concern for others' well-being and a willingness to co-operate for the benefit of everyone – a change in attitude that may be due to having more women in management positions.

The differences in play that Lever found fit into a more general pattern of psychological sex differences described by a number of other researchers. Probably the best known of these is Carol Gilligan of Harvard, whose 1982 book, *In a Different Voice*, sets out a 'psychology of women'. Arguing that most of current psychology is based only on males, Gilligan maintains that females have a very different psychological development and should be studied separately.

In females, Gilligan says, morality and sense of self evolve in a way distinct from males, and the difference is apparent from early childhood on. She describes an experiment in which she asked an eleven-year-old boy and girl how each would solve an ethical dilemma: If a man is too poor to afford to buy a drug needed to save his wife's life, should he steal it? The boy answered yes, the man should steal the drug, because a human life is worth more than the money the drug-store owner would lose. The girl said no, because the man might get caught and go to jail, leaving his wife with no one to take care of her.

The answers illustrate a basic sex difference, Gilligan says. Males tend to make ethical judgments based on an idea of what is right and wrong, very much as they apply rules in a game of baseball to decide if a player has scored a run. In this case, the boy simply applied the axiom that human life is more precious than property. Females, on the other hand, think about how their actions will affect other people and look for options that will result in the best outcome for everyone involved. The girl in Gilligan's experiment suggested that the man could go and talk with the druggist and explain his dilemma and perhaps find some way to buy the drug on a payment plan. She was trying to take into account everyone's interests. 'The moral imperative that emerges repeatedly in interviews with women is an injunction to care, a responsibility to discern and alleviate the "real and recognizable trouble" of this world,' Gilligan writes. 'For men, the moral imperative appears rather as an injunction to respect the rights of others.'

A second, related theme in Gilligan's book is the difference in the way males and females view relationships. Girls and women see relationships in terms of a web, with family, social and emotional ties linking many people into one big network. Boys and men view their relationships with other people – particularly other males – in terms of a hierarchy and always want to know who is on top. Females seek to find common ground and rapport with other people, while males tend to compete with and measure themselves against others.

Thus the psychological and moral growth of the sexes starts from two different points and meets somewhere in the middle, Gilligan says. Females begin by caring about others, often to the exclusion of themselves, and must learn to weigh their own happiness in deliberations about how to act. Males start off focused on themselves and on rules of behavior and must learn to take into account the needs of others. The male and female ethics are opposite but complementary, and well-adjusted adults must incorporate both.

In her popular 1990 book, *You Just Don't Understand: Women and Men in Conversation*, psycholinguist Deborah Tannen showed how the difference in the way men and women view relationships is reflected in the way the two sexes speak. Echoing Gilligan, she wrote that 'women speak and hear a language of connection and intimacy, while men speak and hear a language of status and independence.' For instance, one male/female conversation that Tannen described went like this:

He: I'm really tired. I didn't sleep well last night.
She: I didn't sleep well either. I never do.
He: Why are you trying to belittle me?
She: I'm not! I'm just trying to show I understand!

The woman was only trying to establish connection by showing that she had some of the same problems, but the man interpreted her comments as trying to make his complaint seem less important – as a challenge instead of an offer of sympathy. Tannen says that because the sexes have very different ideas about communication – women see it as a way of establishing connections and men as a way to solve problems and establish rank – people often misunderstand what a member of the opposite sex means.

Anyone who has read Tannen's book has seen himself or herself in it. One night my wife and I were eating dinner with seven friends in a Thai restaurant. At one end of the table were Amy and four other women talking about people they knew; at the opposite end, I sat with the one other adult male and his two sons talking about football. Only an hour before, we had been discussing Tannen's book and, in particular, her observation that males in a group will look anywhere but at each other when talking, while females will look directly into one another's eyes.

Suddenly one of the boys giggled and pointed out what the nine of us had been doing. The women were all leaning into the table with their eyes on whoever was speaking, while we four males were leaned back in our chairs, looking in various directions – at the wall, at the table, at a spot a few inches over someone's head. We males all laughed, but I think it made us a little uncomfortable. We didn't want to go back to staring past each other, and we sure didn't want to start looking into each other's eyes, so we made a big joke out of it. One boy turned completely around to face the opposite wall as he talked, the other covered his eyes with his hand, and I stuck my head under the table. It was easier than eye contact.

Many people read what Gilligan and Tannen write and immediately accept it as true since it resonates so deeply with their everyday experience, yet the work is not universally accepted. One criticism is that these researchers are not doing standard, objective laboratory tests but rather depend on personal observations of small numbers of people – a method that is unprotected from the researcher's biases and expectations and that can give different results depending on which few people you choose to observe. Still, this type of research is the only way to approach things like conversational style. You can hardly give someone a pencil-and-paper test to discover how he or she talks to other people.

The sharpest criticism comes from other feminists who reject Gilligan's and Tannen's depiction of women as having a different but equally valid way of interacting with other people. This debate illustrates a basic division among feminists on how women should be seen. Some continue to insist on the original vision that men and women are inherently the same and only society stands between us and a unisex utopia. Others believe that women are indeed different from men, that those differences should be understood and accepted, but that society must stop thinking of the male approach as the 'standard way' or the 'right way'.

Whatever the reasons for them, psychologists have accumulated a large list of ways that men and women differ in personality and temperament. A standard way to measure these traits is with paper-and-pencil tests that ask a person to agree or disagree with various statements: 'If someone cuts in line in front of me at the store, I'll say something.' 'Physical violence is acceptable in certain situations.' 'When someone crosses me, I'll find a way to get back at him.' If you agree with a series of questions like this, for instance, you'll score high on an aggression scale. Although it might seem as if such self-assessment tests would depend more on a person's self-image than real attributes, psychologists have found that the results of these tests tend to agree well with ratings done by other people, such as friends and family.

In general, females get higher scores on tests of nurturance, emotional responsiveness ('I notice if someone else is upset') and harm avoidance. Males, on the other hand, rate higher on aggression, especially physical aggression, dominance, stimulus-seeking and risk-taking. People high in the last two categories tend to enjoy things like parachuting, hang-gliding and bungee-jumping.

The largest and most consistent psychological sex difference is in aggression. It begins in childhood, when boys are much more likely to take part in play fighting (and real fighting) than girls. Although it is not aimed at hurting anyone, many researchers believe that biology pushes boys to engage in this rough-and-tumble play as practice for real fighting later on. In many mammals, including both rats and monkeys, young males exhibit much more play fighting than young females, and scientists have shown in other species that the behavior is caused by testosterone exposure in the womb.

As adults, neither males nor females are likely to be physically aggressive toward others, but psychological experiments find that, if given the chance, men will be more aggressive than women. In one of the most common experiments a subject is told to take the part of 'teacher' and deliver electric shocks to a 'student' whenever the student gets a wrong answer. The subject presses a button and sees the other person reacting as if shocked. Although there is no electric shock and the 'student' is merely acting, the subject doesn't know this, and the object of the experiment is to see how high the subject will turn up the voltage before stopping. Some 'teachers' take their role so seriously that they will continue to shock a 'student' who appears to be in total agony. In this and a variety of other types of experiments, men have proved to be more aggressive than women, especially when the aggression involves pain or physical injury.

A very practical effect of this sex difference in aggression is the near male monopoly on violent crime. In the United States over the past twenty years women have consistently made up only ten to eleven percent of those arrested for violent crimes – murder, aggravated assault, armed robbery and so on. Although the women's liberation movement has led women to claim a greater share of jobs as doctors, lawyers, scientists and politicians, they haven't increased their share of violent crime. And the types of violent crimes that women commit are different from the male offenses, says Rita Simon of American University in Washington, DC, an expert on women and crime. For instance, about twelve to fourteen percent of the people arrested for murder in the United

States are women, but most of them have killed someone they're close to – husbands, ex-husbands, lovers and sometimes even their children. Usually it's only men who kill people they don't know.

Men and women also differ in competitiveness, although the differences seem to lie not in how competitive they are but rather how they are competitive. In one study of a thousand college students, Brian Gladue at North Dakota State University found no sex difference in what he called 'internal competitiveness', or the desire for personal excellence. But in 'external competitiveness' – the desire to beat somebody into the ground – the men scored much higher than the women. And a study of two hundred college students enrolled in physical education classes found the men to be slightly more competitive than the women in sports, but the major difference was in what motivated them. Women were much more interested in meeting personal goals and standards. Men were more interested in winning.

Of all sex differences, the ones on which we have the most data are the differences in 'cognitive' functioning – how the mind works. Thanks to the American penchant for gauging students with standardized tests at every step from kindergarten to college, it's possible to compare the scores of tens of millions of boys and girls on an array of IQ and achievement tests. As we saw in the last chapter, females tend to outperform males on standardized tests of verbal ability, and males, after about age twelve or thirteen, excel on standardized tests of mathematical ability.

The problem with these standardized tests is that they are not set up to study sex differences – they're intended to estimate scholastic potential or measure how well a child is doing in school. The Scholastic Aptitude Test, for instance, is designed to predict how a student will perform in college and it does a good job of that, but it was never meant to disclose how males and females differ. Even worse, test designers and the companies that publish the standardized tests have worked to minimize the sex differences on many of their tests. So using these standardized tests to gauge sex differences is a little like trying to measure your waist size with a yardstick. You can do it, but there are easier, more accurate ways.

Instead, sex difference researchers have begun relying more and more on specialized tests that show large variations between males and females. In this way they can zero in on the particular mental traits that differ most between the sexes, with the goal of understanding just what the major differences are and why they arise.

Recently, for instance, Melissa Hines at the University of California, Los Angeles reported a very large sex difference in what she calls 'associational fluency'. She recruited fifty-two male and eighty-nine female undergraduates from psychology classes and asked them to come up with synonyms for such common words as *strong, company, turn, clear, sharp, dark, wild* and *tell*. They had one minute for each word. (Try it – it seems easy, but once you get past three or four synonyms it gets harder and harder to come up with more.) When Hines analyzed the results, she found that the women had come up with an average of four synonyms for each word, but the men had averaged only two and one-quarter, or little more than half what the women did. It's not that men have smaller vocabularies – many studies have shown that males and females know the same numbers of words – but rather that the men were somehow less efficient in accessing their vocabularies.

Even Hines was surprised by how big the difference was, and she warns that the results need to be replicated before they can be completely trusted, but the effect seems to be real. Women (at least those of college age) can come up with more synonyms in less time than men. Granted, this may not be a big deal by itself, but the experiment does illustrate a basic principle about the differences between the sexes: they are generally found in very specific skills, such as coming up with synonyms for a given word. In broad tasks that incorporate many abilities, the differences tend to be smaller and harder to notice.

Writing a paragraph, for example, demands a variety of skills – grammar, vocabulary, a sense of structure and so on. Women have an advantage on some of those things – grammar, spelling and thinking up words with just the right meaning – while some of the skills are sex-neutral, and men may even have an advantage on some. Putting it all together, women do better than men on writing tests such as the writing portion of the SAT, but the difference is smaller than the disparities seen in some of the specialized skills.

By focusing on specific abilities such as the associational fluency that Hines tested, researchers are tracing out 'intellectual profiles' of men and women. In language skills, for instance, many of the sex differences lie in fluency – the ability to quickly come up with words or sentences. One example is thinking up synonyms. Another is 'word fluency', the ability to generate words starting with a certain letter. ('You have two minutes to list as many words as you can that start with the letter *d*. Go.') Women consistently do better on these tests, although their superiority on them is much smaller than Hines discovered for synonym searching. A third verbal skill on which women excel is 'expressional fluency', or inventing sentences that fit a given rule. ('Write out as many four-word sentences as you can in which the words start with the following letters: G____ l____ f____ r____. They must be grammatical and have some meaning, and you can't make a second sentence by changing only one, two or three words from an earlier one.')

My own experience with expressional fluency is mostly secondhand – through my wife, who makes up songs and sings them whenever the fancy strikes her. Usually she borrows the melody from another song and invents the words as she goes along – making sure they rhyme and have good meter – but sometimes she makes up the tune and the words at the same time. Once around Christmas she sang what I can only describe as a 'Hannukah carol' – a little song about celebrating Hannukah, complete with candles, menorahs and good food – and I thought it was the real thing. The words were polished, the tune catchy, and she sang it without hesitation or stumbling. 'Where did you learn that?' I asked, puzzled because she's not Jewish. When she told me she'd made it up on the spot, I didn't believe her at first, but since then I've seen her do similar things enough times that I no longer have any doubts.

Females also outperform men on tests of verbal reasoning and comprehension of difficult reading material, and they have a relatively small but predictable advantage on anagrams. ('Unscramble the following to create normal English words: ANTLEG, MAYFIL, MANDRO.') There is no sex difference on vocabulary tests.

Many of the largest and most consistent sex differences in cognition are in the group of skills described as spatial or visual-spatial ability. Spatial ability is somewhat difficult to define, but it is called into play whenever someone must visualize or mentally manipulate geometric objects. An architect drawing a building or an engineer designing a complicated part for an automobile both depend on spatial ability, as does a person locating himself on a map and figuring out how to get home.

Psychologists have developed a wide variety of tests that measure spatial ability. One of the most popular is a mental rotations test, which asks that a subject look at drawings of two three-dimensional objects viewed from different angles and decide if they are identical. The subject must mentally rotate one or both of the objects to see if they match up with one another. In another type of spatial test, a subject must decide what a three-dimensional box will look like after it is opened up and flattened out. Hidden figures tests ask a subject to decide if one simple line drawing is hidden inside another, more complicated figure. And paper folding tests involve mentally punching a hole through a folded-up piece of paper and then figuring out where the holes will be when the paper is unfolded.

If a person does well on one of these tests, he'll likely do well on all the others also, which is why psychologists consider them all to measure one particular ability: spatial ability. By contrast, a person who scores high on verbal tests may or may not get high marks on spatial tests. Verbal ability and spatial ability are independent skills, and a person may have high aptitude for one or the other, both, or neither.

Of all the mental traits that show sex differences, spatial ability causes the fewest arguments. That's not because spatial ability has no practical implications – on the contrary, it's hard to understand how someone could be a good engineer, architect or builder without good spatial skills, and spatial ability also seems to be important in many areas of math and science. But study after study has confirmed that males have a large advantage in spatial ability, and even the critics of sex difference research admit that this difference exists. What arguments remain center on the cause of the difference, but that's a topic for later chapters.

Just how big are the sex differences in spatial ability? Researchers usually talk about the size of a sex difference in terms of a quantity called the effect size, or d. For those familiar with statistics, d is defined as the average male score minus the average female score, divided by the standard deviation. For those not familiar with statistics and who don't want to be, it's easier to think of d simply as a number that measures how big a sex difference is. If $d=0$, there is no sex difference; if d is around 0.2, the difference is usually considered to be small; 0.5 represents a medium-sized difference; and 0.8 and above is large.

The largest effect sizes are for physical attributes such as size or strength, in which men greatly surpass women. For height, $d=2.0$, and the average effect size for how far seventeen-year-old boys and girls can throw a ball is approximately 3.0. When an effect size is this large it means that there is very little 'overlap' between the sexes. If, for instance, you take a group of fifty men and fifty women and divide them into the tallest fifty and the shortest fifty, the tall half would include on average forty-five men and only five women. Conversely, a small effect size means a lot of overlap. Take essay writing, where the effect size is only about 0.09. If you divide the same group of one hundred people into the fifty best essay writers and the fifty worst, on average you would expect the top fifty to contain twenty-six women and twenty-four men.

The effect sizes for mental and psychological attributes are generally much smaller than for physical characteristics. The sex difference that Hines found in coming up with synonyms was one of the largest reported cognitive sex differences, but its effect size was still only 1.2 – less than half that for height. For that reason, some people argue that the sex differences in mental abilities are really quite minor. When looked at from other angles, however, they don't seem minor at all.

In Hines's study, the effect size of 1.2 implies that seventy-three percent of the females but only twenty-seven percent of the males got an above-average score on the test, and in a group of fifty men and fifty women, the top ten synonym finders would include nine women and only one man. If Hines's undergraduate subjects had been taking the test for a grade, you can bet that the males would have thought that there was a large – and unfair – sex difference.

Among other verbal abilities with a sex difference, such as word fluency or anagrams, most have small effect sizes of around 0.2 – big enough to measure, but not big enough to have much practical effect.

Spatial ability is another matter, however. The size of the sex difference here is large enough that it can't be ignored. On tests of mental rotations, for instance, males consistently outscore females, and an effect size of $d=0.8$ is typical. In our hypothetical group of one hundred people, the fifty best at mental rotations would include thirty-three men and seventeen women. The top ten would include eight men and two women. To put it another way, the sex difference in mental rotations ability is about the same size as the difference in IQ between typical college freshmen and people who have Ph.D.s, or the difference in height between thirteen- and eighteen-year-old girls. On other tests of spatial ability besides mental rotation, effect sizes range from as low as 0.2 or 0.3 to as high as 1.0, and all of them favor males.

If you've ever wondered why almost all of the teenagers you see playing video games in a video arcade are boys, the answer may lie in a particular type of spatial skill. 'Dynamic spatial ability' is the knack for watching a moving object and being able to predict its path. Hitting a fastball, firing a shotgun at a clay pigeon or throwing a football to a speedy receiver all require dynamic spatial ability. More than a decade ago, one researcher reported a big male advantage on a primitive type of video tennis game. The male and female subjects began the experiment doing equally poorly, but over the course of five days of practice, the males improved much more quickly even though the females were more conscientious about showing up for practice and were just as interested in getting high scores.

More recently, Douglas Jackson at the University of Western Ontario designed a computer game in which the player has to shoot at objects moving across the screen by using computer keys to aim at the moving targets. The sex difference is even larger than for mental rotations, he says, and it doesn't appear to be due to males having had more practice at video games. When the subjects were first exposed to the game, there was little difference in scores. Instead, like the earlier experiment, the sex difference got larger as the trials went on. If the male advantage was due to previous experience, the sex difference should have shrunk as the women got more practice. Given these results, Jackson says, the reason that most video-game addicts are boys may be found in the old adage that you like what you're good at. Males get more pleasure than females out of games that challenge their dynamic spatial skills.

And it's not just video games where males show an advantage firing at a moving target. In shooting competitions with a moving target – such as trap- or skeet-shooting – men consistently outpoint women. It's not that men are more comfortable with guns, since the sexes do about equally well in such things as small-bore rifle competition, where the target is fixed. Indeed, in rifle competitions sponsored by both the National Collegiate Athletic Association and the National Rifle Association (NRA), men and women compete in one open division on an even basis, and a woman won the 1993 NRA national rifle title. But a sex difference appears when the target starts to move, and in trap-shooting, where the target comes flying out at unpredictable trajectories, most of the top competitors are men.

The male advantage in spatial ability probably has greater practical implications than most other sex differences because it is one of the largest differences and because spatial ability is important in many jobs – and not just for quarterbacks and trap-shooters. Researchers have found, for instance, that the high school students with high spatial ability are the ones who are most successful in geometry, mechanical drawing and shop classes, and in mechanical drawing and shop, spatial ability is actually more important than IQ. The skills learned in these classes are important for careers in science, engineering, drafting and design, and studies have shown that high spatial ability is related to success in such diverse jobs as automotive mechanic, architect and watch repairman.

The sex difference in spatial ability may also spill over into mathematics, where males have a medium-sized edge, larger than the sex difference in verbal abilities but smaller than the difference in spatial skills. (The advantage increases among the very smartest students, as Camilla Benbow has shown.) Since success in many math courses – particularly geometry, trigonometry and calculus – strongly depends on spatial skills, several researchers have suggested that the sex difference in math may be mostly a result of males having better spatial ability. If so, spatial ability will be important in any career in which math is important, which includes many areas of science.

It may be a coincidence, but in researching this book I noticed that many of the female psychologists working in sex differences have a math connection. They either received an undergraduate degree in math, took a lot of math courses in college, or at least said they liked math and were good at it in high school. If nothing else, it illustrates the point that few people, male or female, are successful in science unless they are comfortable with math.

Besides verbal, spatial and mathematical abilities, there are several other mental functions that show sex differences, usually in favor of females. Women are generally faster at extracting details from drawings or strings of letters or numbers, a skill called 'perceptual speed'. One test of perceptual speed is the Number Comparisons test, on which a subject looks at pairs of numbers – such as 20405545 and 20405455 – and must decide as quickly as possible whether the two numbers in each pair are the same. Another is the Identical Pictures test, in which the subject is shown a simple line drawing and must quickly pick out which of five very similar drawings is identical to it.

Females also have better memory than males, at least for certain things. A study by Diane McGuinness with colleagues Amy Olson and Julia Chapman is typical. They tested groups of third-graders and high school students by giving them lists of easily pictured nouns – apple, bird, butterfly and elephant, for instance – and asking them to decide whether each word was more like a male or more like a female. The male/female choice was only a ruse to expose the subjects to the words, however, and after the subjects had handed in their answer sheets with their ratings, each was given a second sheet of paper and told to write down as many of the words as he or she could remember. In each age group the females remembered more words. Among the third-graders, who had been given a list of twenty words, the female edge was about one and a half words on average. For the high school students, who were shown forty words, the females averaged two words more than the males.

The males drew even, however, when the researchers showed the students pictures of an apple, a bird, etc. instead of displaying the words themselves. McGuinness concluded that females have a better short-term memory for words but that males have as good a memory for simple images as females.

In one cross-cultural study, high school students in the United States and Japan were read a short (less than one minute long) story and then asked to write down as many details of it as they could remember. The Japanese students remembered half again as many details as the Americans, but in both countries the girls recalled more than the boys and the female advantage was the same in each country – about twenty percent. In general, although most sex difference studies are performed in the United States and other Western countries, similar sex differences appear when researchers test men and women from other cultures.

The lesson that emerges from all these specialized tests is that males and females have different patterns of mental traits. Males excel in a limited number of areas – various sorts of spatial skills and, to a lesser degree, mathematical ability, which is probably related to spatial ability – but their advantage in these areas is relatively large. Females excel in many more qualities – fluency and various language skills, perceptual speed and some forms of memory – and their superiority in these traits is somewhat smaller.

The same picture – with a few more details – is drawn by studying the patterns of males' and females' scores on standardized tests. For his book, *Bias in Mental Testing*, Arthur Jensen analyzed how the scores of men and women differed on the eleven subtests of the Wechsler Adult Intelligence Scale, applying a technique called factor analysis. Scientists use factor analysis to break intelligence into component parts, on the theory that there are various types of intelligence which come into play in different situations. Jensen found that how well a person did on the Wechsler's subtests depended largely on three separate intelligence factors. The first and most important was the so-called 'general intelligence factor', or g, which can be thought of as a sort of generalized mental efficiency. A person with high g will usually do well in school, score well on standardized tests and typically be recognized by others as being 'smart'. In particular, those people with a high g were likely to score better on the overall Wechsler IQ test than those with low g. Jensen also identified two types of specialized intelligence that influenced performance on the subtests — verbal intelligence and a 'performance' factor, which combined both spatial and quantitative ability. Among two people with equal gs, the one with a higher verbal intelligence would be likely to do better on the subtests that asked verbal questions, while the one with higher performance intelligence would score better on the quantitative and spatial questions.

Jensen found that, on average, women have higher verbal intelligence while men have higher performance intelligence, which is not surprising from what we've already seen. He saw no difference between men and women on g.

But Jensen documented one further difference in the male and female patterns of intelligence that may have great practical importance: males seem to be slightly more variable. In both sexes, most people tend to have intelligence scores somewhere near average, either slightly above or slightly below, and the farther up or down the IQ scale you go, the fewer people you find at that level. The average IQ is 100, for example, and nearly a quarter of the population has IQs in the ten-point range from 95 to 105. There are fewer in the ten-point range from 105 to 115 or from 85 to 95, fewer still between 115 and 125 or between 75 and 85, and so on. If you draw a graph of the distribution of IQs, you get the familiar 'bell-shaped curve'.)

If males' intelligence is more variable than females', there should be a smaller percentage of males than females with IQs between 95 and 105 but a larger percentage of males at both extremes — below 70 and above 130, for instance. In other words, although men and women have the same average intelligence, there would be more men than women who are really smart but also more who are really dumb.

The difference in variability is not large. In statistical terms, the standard deviation is about one IQ point higher in males than in females. You wouldn't notice it in a classroom of normal children, but even a small sex difference in variability, if it exists, would be noticeable at the extremes. A difference as small as one IQ point in the standard deviation would guarantee that nearly twice as many males as females had IQs below 55 — and nearly twice as many had IQs above 145.

For that reason, this is a touchy subject. If it's true that males are more variable, then there would inevitably be more male than female geniuses, a possibility that disturbs many people, and the fact that there would also be more males who are mentally deficient doesn't seem to restore the balance. A number of scientists have questioned the claims of greater male variability, but the evidence does seem to point in that direction.

Few question, for instance, that more males than females are mentally retarded, and the greater number of boys in remedial reading classes may be partly caused by greater variability in reading ability among males. Camilla Benbow finds that among mathematically precocious youth, boys are more variable than girls in their scores on the math section of the SAT. Benbow also points out that boys are much more variable than girls on tests of spelling. As a result, although girls are much better spellers than boys on average (the average boy is worse than seventy percent of girls), the top spellers include equal numbers of males and females. And in a massive study, David Lubinski of Iowa State University and René Dawis of the University of Minnesota studied standardized test scores of 360,000 high school students. They calculated that boys are more variable than girls on every scale: verbal, mathematical, spatial and overall intelligence. (The males' standard deviation was greater by as much as one IQ point.)

In summary, there is no simple way to compare males and females. They vary on a variety of physical, psychological and mental traits, some of them that fit the stereotypes and some that surprise us.

There are two important things to keep in mind when considering these differences. First, in most ways, men and women are much more similar than they are different. It's possible, for instance, to find statistical differences between the sexes on the mathematics portion of the SAT, but if you took a stack of graded SAT answer sheets, covered the names, and asked someone to separate the sheets into a male stack and a female stack, it would not be remotely possible. By focusing too much on the differences, we tend to forget the extent of the similarities.

And second, the existence of differences does not mean that one sex is superior to the other. Carol Gilligan argues in *In a Different Voice* that the moral codes of boys and girls – based respectively on a consideration for self and a consideration for others – are complementary and that a fully adult code is formed by integrating and reconciling the two immature ones. In other words, you need both. In *You Just Don't Understand*, Deborah Tannen concludes that while men and women do have different speaking styles, each is 'valid on its own terms' and neither can be considered as the 'correct' way to communicate.

Diane McGuinness says it particularly well when discussing cognitive differences: 'The sexes do not differ in intelligence but only in the choice of "tools" they employ to solve problems and in the type of problems they choose to solve.' Figuratively speaking, men and women may carry different items in their toolboxes, but the toolboxes are the same size.

CHAPTER 3

Beyond the Birds and the Bees

In 1985, when she travelled to Kobe, Japan to compete in the World University Games, Spanish hurdler Maria Patiño got the shock of her life. Like other female competitors there, she had to take a sex test to prove she was not a man in disguise, but she wasn't worried. She had passed a similar check-up once before, and anyway, she had no doubts that she was a woman. That faith was about to be tested.

Although most people outside the athletic world never hear about sex testing, world-class female athletes know it well. It began at the 1966 European Track and Field Championships in Budapest in response to consistent rumors that some of the best women athletes from the Soviet Union and Eastern Europe were actually men. That year the women at the meet had to parade naked in front of a panel of gynecologists. Although none were disqualified, several of the Communist competitors – including Tamara and Inira Press, who between them had won five gold medals and set twenty-six world records from 1959 to 1965 – failed to show up for the test. This convinced people that the sex test was a good idea.

Two years later, to screen women athletes at the Summer Games in Mexico City, the International Olympic Committee settled on a more dignified method than the 'nude parade'. It was a newly developed procedure called the buccal smear. A lab technician would scrape a few skin cells from the inside of a woman's cheek, then stain the tissue sample and examine it under a high-powered microscope. If the cells had two X chromosomes – the mark of a genetic female – then the technician would see a dark spot inside the cell's nucleus called a Barr body. If only one X was present, as is the case with genetic males, then no dark spot appeared.

It was this test that Maria Patiño took in 1985 to prove she had no unfair advantage over the other women hurdlers. When her results came back, however, there were no Barr bodies in her cells – genetically she was a male, the test said. Meet officials told her she would not be allowed to compete, so Patiño faked an injury and left. But convinced that she was just as female as the other competitors, she continued training and entered a meet in Spain several months later. Ignoring a warning not to compete from the president of the Spanish athletic federation, she won her event, but the following week she was kicked off the Spanish national team, stripped of her titles and barred from competition.

Is Maria Patiño the woman that she appears to be, or the man that the buccal smear says she is? She is not a transsexual – that is, someone who has undergone surgery to change from one sex to another. She has been a female all her life. But nor was there a mistake with the buccal smear – she does not have two X chromosomes. So what is she (or he)? The answer takes us to the heart of what makes a male or female.

Sex seemed so simple back in ninth-grade biology class, when the teacher explained that it all depends on the X and Y chromosomes. Now of course I realize that in ninth grade you were probably too busy making your own observations about the opposite sex to listen to much of what the teacher said, but if you had paid attention you would have heard something like this: Every human has twenty-three pairs of chromosomes in each cell of his or her body, and those chromosomes carry all of the genetic information that describes that person, from hair color to hat size. Twenty-two of those pairs of chromosomes are the same in men and women, but the twenty-third is not. These are the sex chromosomes. A woman has a matched set of two X chromosomes, which appear long and elegant when seen under a high-powered microscope. A man has only a single X escorted by a short, squat Y chromosome. This is the pattern among all mammals, from chipmunks to the blue whale: females, XX, and males, XY.

But for Maria Patiño the simple answer just won't do. She has one X and one Y chromosome, not two Xs, yet to all appearances she is a female. To understand her sex, we must go past the chromosomes to see how a body develops in the womb.

A human embryo is neither male nor female for the first seven weeks after conception. It is essentially neuter, and it is impossible to tell an XX fetus from an XY fetus without peering deep into the individual cells and examining the sex chromosomes themselves. At seven weeks, the fetus has an embryonic reproductive system consisting of a pair of gonads which can grow into either ovaries or testes, plus a mass of tissue called the genital ridge which can develop into either a clitoris and labia or a penis and scrotum. There are also two primordial systems of ducts, one female and one male. The female ducts, called Müllerian ducts, grow into the uterus, fallopian tubes and part of the vagina if the fetus heads down a female path, while the male, or Wolffian, ducts are the precursors of the usual male plumbing – the seminal vesicles, vas deferens and epididymis.

During the eighth week the fetus chooses between two paths: masculine or feminine. If the fetus is genetically male – if it has a Y chromosome – then a 'master switch' on the Y chromosome clicks on at this time. This switch, which is a single gene called the testis-determining factor, triggers a whole series of events that will point the fetus in a male direction and culminate in a baby boy – if everything goes as planned.

The flipping of this switch is the key step. The testis-determining factor signals the embryonic gonads to form into testes, which then begin to produce male hormones, and the hormones take it from there. The major hormone produced by the testes is testosterone, which stimulates the Wolffian ducts to start developing into the male duct system. At the same time, some of the testosterone is converted by the body into a second male hormone, dihydrotestosterone, which prompts the genital ridge to begin forming into male genitals. The testes also produce a substance called Müllerian inhibiting factor, which causes the (feminine) Müllerian ducts to atrophy and eventually be absorbed by the body.

A female fetus, on the other hand, doesn't have a Y chromosome and so the master switch is never turned on. Nothing happens in the eighth week. Instead, the fetus continues to grow and develop, and in the thirteenth week the gonads start to transform into ovaries. Without the presence of the testes to turn out large amounts of male hormones, the rest of the sexual system, internal and external, develops in a female direction.

In other words, the 'default' body plan is female, and a fetus will go down this path unless diverted. The gonads will become ovaries unless the testis-determining factor tells them to transform into testes. The genital ridge naturally evolves into the clitoris and labia, and the Müllerian ducts mature into the uterus, fallopian tubes and upper one-third of the vagina, unless they are disrupted by male hormones. The Wolffian ducts will shrivel up without the intervention of those same hormones.

So sex is not so much a matter of chromosomes as of hormones. Indeed, the small Y chromosome, the root of all maleness, seems to do little besides turn on the 'master male switch' to start the flow of hormones. The Y holds relatively few genes and most of them are duplicates of genes that lie on the X chromosome – as far as researchers know, there are only a handful of genes that are unique to the Y chromosome. These include the testis-determining factor, a gene involved with male fertility and making sperm, and a third called the H-Y antigen gene, which has nothing to do with sexual differentiation, as far as researchers can tell. In short, the Y chromosome is little more than an abbreviated version of the X with the gene for the male switch tacked on.

In practical terms, this arrangement means that it is the hormone environment of the womb, not the chromosomes, that directly determines the sex of the fetus – a fact that can be demonstrated dramatically with a simple lab experiment. Take a female rat early in her pregnancy and inject her with a large amount of testosterone. When her pups are born, they will all be male. Or more accurately, they will all look like males even though some have XX chromosomes and others XY. The testosterone has masculinized them all, even the ones who otherwise would have become female. Only the internal sexual organs set apart the XY sons from the XX 'sons' – the XX rats will not have testes because they had no Y chromosome to turn the gonads in that direction, and the Müllerian ducts will have evolved into uterus and fallopian tubes because there was no Müllerian inhibiting substance to stop them.

This is basically what happens to the spotted hyena described in the introduction to this book. High levels of testosterone in the mother's womb create masculine-looking genitals on her female pups, although their internal sexual organs are not affected. The process is more complicated than simply adding extra testosterone since the female hyena had to evolve ways to have sex and deliver babies with her seemingly male equipment, but the basic principle is the same as with the laboratory rats.

It is also possible to perform the reverse experiment on rats. By injecting the mothers with a substance that prevents testosterone from doing its masculinizing work, researchers can produce a litter that looks all female, although some of the pups have a Y chromosome. Because these XY females have testes which produce the Müllerian inhibiting factor, they never develop the internal female organs, but externally everything is completely female.

And in essence, this is what happened to Maria Patiño. In humans, approximately one of every twenty thousand genetic males has a defect in his 'androgen receptors' – the large molecules that act as middlemen between male hormones (androgens) and the various tissues in the body that these hormones act on. When such receptors are working right, they grab hold of molecules of male hormones and then bind to the DNA in the cells, triggering a variety of biochemical events that eventually lead to such things as the growth of the penis and scrotum. The hormones cannot turn on these events by themselves – they need the receptors, and if the receptors are defective the body will not respond to testosterone and the other male hormones.

In such cases of 'androgen insensitivity', the XY fetus will develop testes around the eighth week of gestation, right on schedule, and soon afterward the testes start producing testosterone and sending it out into the body. But as far as the rest of the body is concerned, the testosterone isn't there. Because the receptors are faulty, the body cannot detect any male hormones, so the genetically male fetus never hears its 'male call' and instead heads down a mostly female track. The genital tissue shapes itself into a clitoris and labia, and the male Wolffian ducts shrivel away. The testes still produce Müllerian inhibiting factor, however, so the Müllerian ducts atrophy and the fetus never develops a uterus, fallopian tubes or the upper part of the vagina.

At birth, the child looks like a girl. The only sign that something is different is the presence of the testes, either in the labia or in the lower part of the groin, but doctors will miss those without a close examination. At puberty, the girl develops breasts and a woman's body. Although she has no ovaries to make female hormones, her body transforms some of the testosterone and androstenedione produced by the testes into the female hormone estradiol, and there is enough estradiol to trigger her maturation into a woman. Because she has no uterus, however, she does not menstruate. This may prompt her to go to a gynecologist, who will discover that she has no internal female organs, her vagina has a dead end, and she is carrying a pair of testes. The doctor will usually recommend removing the testes because they often turn cancerous, and then will prescribe estrogen replacement to substitute for the hormones that had been produced by the testes. If, as sometimes happens, the vagina is too short for comfortable intercourse, it can be stretched.

Other than that – and the fact that she cannot have children – the XY woman is no different from an XX female. If anything, she's more likely to match the Western male standards of a beautiful woman, with long legs, well-developed breasts and clear skin. She may become a fashion model, an occupation that supposedly holds a number of XY women, or a movie actress. Indeed, there are at least two well-known American movie stars who are XY women, according to researchers in sex differences, although neither of the actresses wishes her condition to be made public.

The XY woman's height also gives her an advantage in sports, and they seem to be better represented among the top women athletes than would be expected by their one-in-twenty-thousand presence in the general population. Although the exact numbers are hard to come by since the athletic governing bodies don't release them, it's not unusual for several women to be disqualified for failing the sex test at a major international competition, says Alison Carlson, a former Stanford University tennis player who has been active in the campaign to get rid of the genetic testing. In a survey of small competitions, Carlson found that about one in every five hundred women was disqualified, most of them for being XY females.

XY women such as Patiño are by no means the only exceptions to the old rule from biology class that XY equals male and XX equals female. About one in every twenty-five hundred females, for instance, has only a single X chromosome, which leads to a condition called Turner's syndrome. In the womb, these females' embryonic gonads fail to mature into either ovaries or testes, so the fetus has little exposure to either female or male hormones, with the result that the fetus develops as a female. Since they have no ovaries, the girls need hormone treatments to trigger a normal sexual maturation at puberty. Generally Turner's women are quite short, exaggeratedly feminine, and often express a great desire to get married and raise a family. Unfortunately for them, without ovaries, they can never have children of their own.

Some women have an X and a Y chromosome, but the Y chromosome is missing the small part that contains the gene for the 'master male switch', or testis-determining factor. Without this gene, the Y chromosome cannot perform its male-making function, and in such cases the people develop much like Turner's women.

Then there are males who have two X chromosomes, but through some sort of genetic defect the gene for the testis-determining factor has migrated to one of the X chromosomes, where it doesn't belong. This gene triggers the fetus to develop as a normal male, even though it has a female set of chromosomes.

And there are a variety of ways that a person can have one, two or even three extra sex chromosomes – doctors have discovered people with XXX, XXY, XXXY, XYY, XYYY and XYYYY, for example. Surprisingly, a female can have an extra X chromosome (XXX) or a male can have an extra Y chromosome (XYY) and still have completely normal development, although XXY and XXXY individuals tend to have ambiguous genitals.

Much simpler things can go wrong, however, than these rare

chromosomal goofs. One of the most common is for the fetus to be exposed to abnormally high levels of testosterone, much as happens to the spotted hyena.

One child in every fourteen thousand born in the United States has a genetic defect that causes large amounts of male hormones to accumulate in the fetus as it is growing. The disease, called congenital adrenal hyperplasia, or CAH, prompts the body to make too much androstenedione – the same chemical produced in such large quantities by the mother spotted hyena – and this in turn is transformed into testosterone, as it is in the hyena. A CAH baby isn't exposed to as much testosterone as a fetal hyena gets, but there is often more than enough to unsettle the sexual development if the child is female.

In the worst case, a baby girl with CAH will be born with external genitals nearly indistinguishable from the penis and scrotum of a baby boy. Her clitoris has grown to the size of a penis, and her labia have fused to resemble a scrotum. In milder cases, her genitals may merely seem ambiguous – the clitoris may be closer in size to a small penis than a normal clitoris, for instance, but the labia will not have fused. Either way, the internal organs remain completely female. The gonads have become ovaries; the Müllerian ducts have matured into the uterus, fallopian tubes and upper part of the vagina since there was no Müllerian inhibiting factor to hold them back; and the male Wolffian ducts never developed, apparently because there wasn't enough male hormone in the right place at the right time to trigger their growth.

Even if she is born with very masculinized genitals, a CAH girl can live a normal female life with the help of medical treatments developed over the past forty years. Shortly after birth she will be placed on long-term hormone replacement therapy that counteracts the effects of the excess testosterone, and her genitals will be surgically reconstructed into the female pattern. Given this, she should go through puberty at the same time as her female peers and have no trouble getting pregnant or having children. She may be more muscular than an average woman, thanks to the testosterone exposure in the womb, but it's nothing that falls outside the normal female range. In short, although there may have been some initial questions about her sex at birth because of her masculine-seeming genitals, a CAH girl is unquestionably female, her sex matches her chromosomes, and she grows up knowing which sex she is. That makes her much luckier than victims of another genetic disorder, which can leave them confused about just what sex they are.

Physiologist and writer Jared Diamond tells the story of Barbara, 'an apparently normal girl enjoying a happy childhood.' When she was fourteen, however, Barbara started to worry because, unlike her friends, she had not begun to menstruate and her breasts had not begun to grow. Her concerns grew as she discovered a lump in her labia and later found her voice deepening, facial hair sprouting and her clitoris becoming larger and starting to resemble a penis. 'After Barbara's sixteenth birthday, her penis developed erections, she produced ejaculations, and she found herself feeling a sexual interest in girls. By now she had become convinced that she was really a boy and that the mysterious mass within her labia was a testis. But Barbara still struggled with the problem of how to present herself to her parents and friends, before whom she avoided being caught naked. When they found out, would they ridicule her – or him – as a freak?'

Barbara actually was a male. He had XY chromosomes, a penis and testicles, and the normal male interior plumbing. What he didn't have was the enzyme that converts testosterone to dihydrotestosterone, and that one deficiency was enough to keep him looking like a girl for fifteen years. In the womb, the presence of a Y chromosome had caused Barbara's gonads to become testes. As in normal males, they produced both testosterone, which triggered the growth of the internal male organs, and Müllerian inhibiting substance, which kept the internal female organs from developing. But without the missing enzyme, called 5α-reductase, there was no dihydrotestosterone to tell the genital tissue to form into a penis and scrotum and so the tissue grew to look like a clitoris and labia instead. There was even a vagina, although it was a dead end because it had nothing to meet up with.

As a result, the newborn Barbara was identified as a girl. At puberty, however, when his testes began to produce large amounts of testosterone, this hormone completed the job left undone in the womb. His 'clitoris' enlarged to become a penis, his 'labia' grew to form a scrotum, and his testicles – which had been hiding in his groin and one of which had moved into the labia – dropped down into the scrotal sack.

This 5α-reductase deficiency is an extremely unusual condition because it depends upon a child inheriting the same rare defective gene from both father and mother. Most of the cases have been discovered in small, isolated villages in undeveloped countries, where the population is inbred and the chances of both parents having the defective gene are high. In 1979, for instance, a group of researchers led by Julianne

Imperato-McGinley of Cornell University Medical College described finding thirty-eight cases in twenty-three interrelated families living in three villages in one small Caribbean country.

By the time the villagers became aware of the problem and learned to recognize it at birth by the child's somewhat ambiguous genitals, Imperato-McGinley reported, they had raised eighteen of the males as girls. One of them never accepted his real sex and continued to live as a female, but the other seventeen switched to a male identity after puberty and most have had been married or at least lived with a woman. Their sexual organs are adequate for intercourse, but they cannot get a woman pregnant because they ejaculate through a small opening between the legs next to the dead-end vagina. That hole is the exit point for the urethra which normally, of course, would go through the penis, but since the penis never formed in the womb, the urethra ended up close to the normal female position.

In his tale about Barbara, Diamond doesn't reveal the ending, but if Barbara was typical he eventually accepted his maleness, although perhaps with a great deal of counseling and psychiatric help.

Much of what we know about the hormonal control of sex pertains to the body and, especially, the sex organs. Does a fetus develop a penis or a vagina? The hormones decide. But it's also clear – in animals, at least – that the hormonal influence does not stop there. Hormones also determine whether a brain will become masculine or feminine.

If you stop to think about it, it's hard to see how it could be any other way. Rats, for instance, depend on instinct for a great deal of their behavior, including sexual behavior. They don't learn about the birds and the bees from their folks, and they can't look up anything they don't know in *The Joy of Sex*. So somewhere in the rat brain must be some sort of sex circuits that tell the rat what to do when the time is right, and these circuits must be different in males and females. Imagine, for instance, a female rat that thinks she's a male. Even though she has female sexual organs, she would try to mount other female rats and would reject the advances of the males. The species wouldn't survive long like that.

Actually, you don't have to imagine this confused-rat scenario if you're willing to trek out to Roger Gorski's lab at the University of California at Los Angeles. Gorski, one of the pioneers in studying the effects of hormones on the brain, has made a short film to show just how mixed up things can get if brains don't get the right hormones. If they gave out awards for best comedy lab films, Gorski would be an Oscar winner.

The movie has six episodes, each starring two rats. The genetic sex of each rat is easily distinguished because the females are pure white and the males are white with black heads. In the first scene, we see a male put into a box with a female that has been primed with female hormones so that she's ready for sex. (Normally female rats come into and out of estrus – the period of sexual receptivity, like heat in dogs or cats – on a regular cycle, but estrus can be artificially induced by injecting the rat with estrogen and progesterone.) Neither of the rats is very big on foreplay. They just get right to it. The male climbs up on the back of the female, wrapping his forelegs halfway around what would be her waist if she had a waist. In response, the female moves her tail out of the way, tilts her head backwards and arches her back to make it easier for the male to enter her. This instinctual movement is called 'lordosis' and it is a distinguishing feature of the female rat brain – a female does it automatically whenever a male mounts her if she has been stimulated with hormones, either natural or artificial. A normal male will only rarely do it even if given the same female hormones. In the film, after the male rat has mounted, he thrusts a few times and it's all over.

Scene two features a normal male with an outwardly normal female, but this female was treated with testosterone around birth to defeminize her brain. Again the female has been given female hormones, but this time they don't work. When the male climbs up on her back she simply walks away, looking bored. For a while he tries to stay on top of her, trying to hold her waist with his front paws and scrambling awkwardly with his rear feet to keep up. Eventually he gives up, puzzled.

The third episode is back to normal rat sex, but hold on – both rats are females. The one on top, Gorski explains, was treated at birth with testosterone, which masculinized her brain, and then given more testosterone shortly before being put in front of the camera, which caused her to behave like a male when presented with a provocative female.

Scene four is back to a male and female. Again the female was treated with male hormones at birth and given a shot of testosterone just before being put in the box, and now she thinks she is a big male rat stud. When the male attempts to mount her, she slips out of his grasp, pivots, and climbs on top of him, thrusting. Startled, the male runs out from under her, then circles around and tries a side approach, but again she slips away and tries to mount him instead. And on it goes, two rats each following the call of male hormones.

Next are two males, one normal and the other castrated at birth and treated with female hormones before being put in the box. Now everything goes as planned. The first male mounts the second, who stays in place, arching his back and trying to be as receptive as possible, given that he's missing the proper female sexual equipment.

Finally, Gorski shows a male and a female, both confused. The male, castrated at birth and given female hormones before the experiment, takes the part of the female; the female, given testosterone at birth and again just before being put in the box, performs the male role. The scene is indistinguishable from the first, except now it is the female climbing up on the other's back and wrapping her forelegs around his waist, while the male is arching his back and moving his tail obligingly out of the way so that the female can thrust with the penis she doesn't have into the vagina that is nowhere to be found on him.

'The brain is part of the reproductive system,' Gorski says, and just as a rat has either male or female sexual organs, it has a male or female brain, with the sex of the brain determined by the sex hormones it is exposed to. The rat, he says, is a particularly good animal in which to study this hormone action, for two reasons. First, the animal's sexual behavior is very clearly either male or female so it's easy to see how certain hormones have affected it. And second, a rat's brain is still developing at the time it is born. It is not until the first few days after birth that a rat brain settles into being male or female, so the 'brain sex' can be modified by hormone treatments of the newborn rat after it leaves the womb. Castrating a newborn male rat will cause its brain to develop in a feminine direction, while treating a newborn female with testosterone will give her a masculinized brain.

In other words, the brain follows the same basic developmental pattern as the body: left alone, it becomes female; exposed to male hormones, it becomes male.

That's the traditional view – and it is certainly true in general – but recently researchers have started to think that there may be more to a brain being female than simply being not-male. In particular, it may be that female hormones, such as estrogen, play a vital role in making the brain fully feminine. For instance, Dominique Toran-Allerand at Columbia University in New York City has found that estrogen boosts neuron growth in brain tissue from female rats. Without the estrogen produced by the ovaries, she suggests, the brain may be incompletely female or perhaps in some ways even more 'neuter' than female. A testosterone-free environment may be all that's needed for the female reproductive system to develop, she says, but femininity is a more active attribute when it comes to the brain.

There is one further complication to this basic picture of how the brain becomes masculine or feminine. It turns out that much of the work of masculinizing the brain is not actually done by testosterone itself but instead is carried out by estrogen, which is produced from testosterone by the brain. Thus not only can researchers masculinize a female rat's behavior by exposing it to testosterone, but they can obtain a similar effect by treating it with estrogen. So why doesn't the estrogen produced by a female's ovaries masculinize her brain? The answer seems to be that the ovaries don't produce enough estrogen. Combining this fact with Toran-Allerand's work, it may be that the brain needs a certain minimum amount of estrogen to become feminized and then past some point, more estrogen or extra testosterone will act to masculinize the brain.

All this is well-accepted for rat brains, but what about humans? Do sex hormones in the womb mold the human brain at the same time they are shaping the body? The answer is much more complicated – and much more controversial – for humans than for rats, and we'll spend most of the rest of the book looking at the evidence. For now, however, let's examine the proof for a less contentious point. Assuming that there are biological differences between men's and women's brains, then it is the hormones, not the chromosomes, that are the likely culprits.

Suppose, for instance, that there is some biological reason why men do better on tests of spatial skill than women. Could the chromosomes cause this? They might if, for example, the Y chromosome contains a 'spatial gene' that gives a person extra spatial skills. Because it's on the Y chromosome, men would get the gene and women wouldn't. This hypothesis is easy to test since it implies that a male's spatial ability is inherited from his father (because a son's Y chromosome must come from the father). Genetics studies show, however, that spatial ability doesn't work this way – there is no father-son pattern of spatial ability and thus no spatial gene on the Y chromosome.

Indeed, as we saw, there are few genes that lie on the Y chromosome and even fewer that aren't also on the X chromosome. Thus males are unlikely to differ from females in mental abilities because of genes on the Y.

The other possibility is genes on the X chromosome. These could create sex differences because women, having two X chromosomes, have two copies of such genes – one from each parent – while men have only one copy, inherited from their mothers. This is exactly why a sex difference exists in color vision, with more men than women being color-blind. The gene for red/green vision lies on the X chromosome and if a man has a defective gene for red/green color vision on his single X chromosome, he won't be able to tell those colors apart. However, if a woman has such a defective gene on one X chromosome, she'll be OK as long as the color-vision gene on the other X chromosome is functional. A woman needs to inherit a defective color-vision gene from both her mother and her father if she is to be color-blind.

In the 1960s and 1970s, researchers looked to see whether such a genetic arrangement could explain the male advantage in spatial skill. Perhaps more males have high spatial ability for the same reason that more males are color-blind: a recessive gene on the X chromosome. ('Recessive' means that the gene has no effect if there is a second, 'dominant' gene that overrides it.) But after two decades of genetics studies, however, the consensus is that no such spatial gene exists on the X chromosome. Scientists have also looked for other genes, such as a 'verbal gene', on the X chromosome, and so far no one has found evidence of a gene on a sex chromosome that controls any mental or behavioral difference between men and women.

In short, looking for the differences between men and women on the sex chromosomes has proved to be a dead end. If there are biological reasons for any of the differences between men and women, they are almost certainly caused by different hormone environments in the womb.

The question of brain differences aside, there is no doubt that hormones determine which sex a body becomes. That concept alone, once it is widely understood, may change the way people think about masculinity and femininity.

Thomas Laqueur writes in his book *Making Sex: Body and Gender from the Greeks to Freud* that the most popular view for two thousands years – up until the eighteenth century – was that females are merely imperfect males. In this vision of the sexes, male and female genitals were basically identical, but female bodies did not have enough 'vital heat' to push the genitals out. The ovaries were the same as testicles, the uterus was simply an interior scrotum, and the vagina was nothing more than an inverted penis, with the foreskin forming the labia. Indeed, Laqueur says, the same word was used in most languages for both ovaries and testicles, and neither Latin nor Greek nor any common European language before about 1700 had a separate word for vagina.

Laqueur tells a story from the sixteenth century that illustrates this view of the sexes: while in early puberty, a girl named Marie was chasing a group of pigs through a wheat field and jumped over a ditch. This exertion raised her internal heat so much that it transformed her into a male, pushing her vagina out to become a penis and popping out the ovaries to take their place as testicles. She/he later grew a beard, changed her/his name and got a position with the king of France. From what we know now, it's possible that Marie suffered from 5α-reductase deficiency and that she was a boy all along whose male genitals waited until puberty to appear, but it's not important whether the story was real or apocryphal. The point is that for much of recorded history, the sexes were thought of as one basic pattern expressed perfectly in men and imperfectly in women.

That view changed sharply around the end of the eighteenth century, Laqueur writes. For a variety of social and political reasons, people began to think of males and females as completely dissimilar and incommensurate. In place of a focus on the similarities (real or imagined) between the sexes, suddenly the differences became paramount. Males and females became 'opposite' sexes.

As an illustration of this transition Laqueur points to a change in attitude toward female orgasm. When men and women were considered to be different only in degree (how close to metaphysical perfection they were), it was taken for granted that conception could not take place without a female orgasm. After all, a male orgasm was certainly necessary to conceive a child. But after the advent of the two-sex world view, female orgasm was assumed to be irrelevant to conception. This belief (true, as it turned out) was not based on newly found scientific evidence, but arose instead from the practice of assuming the sexes to be different in every possible way.

How should we now view the sexes? Let's start with the facts. The scientific bottom line is that we'd all have female bodies if it weren't for the male hormones produced by about half of us in the womb. Take away the testosterone and it doesn't matter what the chromosomes are, a person will grow up looking like a female.

Now if you wish, you can transform this objective biological fact into some mystical interpretation of what it means to be a male or female. You can argue, for instance, that the female body is the natural human form and that men are therefore less natural than women. It certainly seems that jerking the fetus from its 'natural course' and making it a male has a cost – the extra steps necessary to become a male may explain part or all of the greater number of male fetuses that are spontaneously aborted. Or you might opine instead that the female's two X chromosomes are boring and passive, while the male Y chromosome is a dominating, risk-taking kind of guy. When Mr Y is on the scene, he directs everything, and it's only when he's not around that a female can come to the fore. This may be a good subject to introduce if you want to break up a party that's gone on too long, but you might as well debate how many angels can dance on the head of a pin or which came first, the chicken or the egg.

Neither do the biological facts imply that either of the historical interpretations of sex that Laqueur described is accurate. There's nothing in the hormone picture of sex to imply that the sexes are 'opposite'. Given that a seven-week-old fetus can head down either a male or female path, depending on the hormones in the womb, all humans have the potential for both masculinity and femininity, at least at the very beginning. On the other hand, the older view of sex is equally flawed since it imagined humans to be inherently of one sex – male – and to differ only in how closely they approached that ideal state.

A simpler and less value-laden way to think of male and female is simply as two manifestations, two alternative versions, of a human being. In the womb, a switch is thrown. It is either flipped to the male position, or left in the female position, and depending on that setting the body and the brain will develop in certain characteristic ways. Most of the time the switch works as planned, but not always. Sometimes the chromosomes say one thing, but the switch gets set in the opposite position – as is the case with XY females. And sometimes, as happens with CAH girls born with ambiguous genitals, the switch gets stuck somewhere between male and female.

All this means that the job of defining male and female can sometimes be more complicated than simply asking about chromosomes. Consider an XX rat exposed to testosterone as a fetus so that it has normal male genitalia and is sexually attracted to female rats – is this a female or a male? Or consider the female spotted hyena that has a penis which gets erections, that is larger and more aggressive than a male, but that gives birth to pups and then nurses them – is she any less of a female because she was exposed to high levels of male hormones while in the womb? These are not merely academic questions, as Maria Patiño discovered.

The finish to Patiño's story has some good and some sad. She discovered, for instance, why other female athletes who fail the sex test have traditionally dropped out of competition, faking a career-ending injury and keeping the test results secret. Before her disqualification from the meet, Patiño had been dating several men in Spain, but after it hit the newspapers that she had a Y chromosome, they suddenly stopped calling.

There can be few things more humiliating for a woman than being publicly proclaimed a male, but Patiño accepted that in order to fight what she thought was an unfair position. The International Olympic Committee claims to be protecting women athletes against unfair competition from men disguised as women or from women with genetic abnormalities that give them masculine characteristics, but that doesn't justify barring women like Patiño from competition. Although she does have the X and Y chromosome of a male, this gives her no advantage over the competition. If anything, she may be at a slight disadvantage because her body does not respond at all to male hormones, including the illegal steroids that some athletes use to get an edge over the rest of the field.

In essence Patiño insisted that the sports world defend its answer to the question, What is a woman? The International Olympic Committee, which oversees the Olympic Games, and the International Amateur Athletic Foundation, which runs many other amateur events, had decided that the answer to that question was 'someone who has two X chromosomes'. But Patiño had a different answer: '*I* am a woman.'

In her fight with the Olympic Committee she was supported by a coalition of athletes, doctors and scientists who argued that the chromosome test was the wrong criterion for deciding whether a competitor was female. It would be simpler and more fair to go back to the original idea of checking a competitor's anatomy to make sure that she doesn't have a penis, says Alison Carlson, who has been active in the effort to change the sex-test rules. The real goal of a test should be to make sure that no men disguise themselves to compete in women's events, not to discriminate against people like Maria Patiño who are truly female but who simply got that way by a different route than most women.

That argument is gaining ground. In 1991, the International Amateur Athletic Foundation decided to do away with the chromosome test and to rely on a visual examination. The International Olympic Committee so far has not backed down from its testing position, but it has agreed to recognize sex certificates issued by the IAAF. This opened the door for Patiño, who was certified as a female by the IAAF and was eligible for the

1992 Summer Olympics. But after her two-and-a-half-year battle to get back into sports, and after setting a Spanish women's record for the sixty-meter hurdles, Patiño had to watch from the sidelines. Another woman had overtaken her as the fastest hurdler in Spain, and she edged out Patiño for the spot on the Spanish Olympic team. Still, Patiño would much rather lose to a competitor in a fair race than to never get a chance to race at all. And besides, she had already won the more important contest, which was to get the sports world to acknowledge what she already knew: that she is indeed a woman.

CHAPTER 4

Echoes of the Womb

So many toys to play with and so little time. As psychologist Sheri Berenbaum watches, the little girl surveys the collection of toys in front of her and ponders where to start. Finally she picks up a fire truck and pushes it along the ground, making 'vroom, vroom' noises and a child's imitation of a siren. When she tires of the truck, she reaches for a box of Lincoln logs and builds a small house. Nothing unusual here, it seems – just a normal five-year-old at play. But the child's twelve-minute toy session, videotaped for later analysis, is actually part of one of the most remarkable discoveries about human nature made over the past decade.

Every parent knows that little boys and girls like to play with different toys. Boys choose cars, trucks, building blocks and plastic guns, especially those that actually discharge something (darts are good, water even better). Girls prefer dolls and other 'feminine' things, such as kitchen toys. And most parents will tell you that these preferences appear without any prompting from them. My next-door neighbors, for instance, are liberal, liberated parents who have given their boy dolls and a vacuum-cleaner and their girl trucks and airplanes, but still the boy likes his little cars best, while his sister loves to 'cook'. Whenever my wife visits, the girl will run back to her play kitchen and fix something special for Mommy's guest. (The parents have never given toy guns to either child, but the little boy picks up sticks and tubes of any sort and pretends to shoot them.)

This is typical, Berenbaum says. Many of her friends grew up in the rebellious sixties and swore that their children would never be indoctrinated with the sexual stereotypes of previous generations. It didn't do any good. 'I have one friend who swears there's a truck gene for boys,' she says with a laugh.

Why should this sex difference persist in the face of the best efforts of modern, liberal-minded parents? It's unlikely that boys' brains have some sort of 'truck circuit' whose neurons are arranged differently from a 'doll circuit' in girls's brains, particularly since such toys are a relatively recent cultural invention. Fifty thousand years ago, when our ancestors were anatomically identical to modern man, Stone Age dads were not chipping dump trucks out of flint for their sons' birthdays. The most Junior could hope for was a couple of arrowheads and a pet rock. So, given how slowly evolution proceeds, there has been no time for genetic toy preferences to have evolved.

For this reason, most people assume that the sex difference in the types of toys children prefer must be taught. Even though Mom and Dad may be careful to give dolls to Junior and cars to Sis, they can't keep the kids away from TV or from their peers at school, and sooner or later the children are going to learn which toys are 'appropriate' for their sex. A few hours of watching Saturday-morning cartoons and the commercials that come with them is enough to undo all of a parent's careful non-sexist toy selections. You can damn this as cultural oppression or approve of it as society teaching children their proper roles, but either way it would seem that the boy/truck, girl/doll connection is nothing more than a learned behavior.

That argument is quite convincing and – if Berenbaum's experiment is to be believed – quite wrong. Berenbaum, a psychologist at the Chicago Medical School, studied a group of two dozen 'special' girls, ages three to eight, by putting them in a small enclosure with a collection of toys and then videotaping their play. The girls had a choice of traditional boys' toys (cars, trucks, airplanes and building blocks), traditional girls' toys (dolls and kitchen toys), and neutral toys (board games, puzzles and books). For the sake of comparison, Berenbaum performed the same test with a group of boys and a group of 'normal' girls.

When she analyzed the tapes and added up the time that the children spent playing with different toys, she found that the girls in the first group – the special girls who were the subject of the experiment – acted almost like boys instead of normal girls. They played with boys' toys more than twice as often as with girls' toys, the same proportion that Berenbaum saw among the boys. The group of normal girls, on the other hand, much preferred the girls' toys. All three groups played with the neutral toys for about the same amount of time.

Who are these special girls? Are they the daughters of parents who have found the secret to non-sexist rearing? Were they raised in remote cabins, protected from television and all the other ways that society teaches children how to act their sex? Not at all. The parents of the girls in Berenbaum's study say they did nothing special. They raised their daughters as normal females.

What is different about these girls is that they were all victims of congenital adrenal hyperplasia, or CAH, the condition we met in the last chapter that causes a fetus to be exposed to very high levels of testosterone in the womb. After testing and rejecting other possible explanations for their behavior, Berenbaum has concluded that these girls' preference for trucks and Lincoln logs over dolls and toy blenders must have been caused by something male hormones did to their developing brains.

It takes a little getting used to, Berenbaum admits. The idea that hormones during development could shape toy preferences years later seems strange, almost spooky. But the experimental facts are hard to deny and, furthermore, they are just part of a large and growing body of scientific work with similar conclusions. Researchers have studied a number of different types of people who, like CAH girls, were exposed to unusual hormone environments in the womb, and the evidence all points in the same direction: sex hormones during development modify the brain, influencing a variety of mental and psychological traits later in life. Still, Berenbaum says, she sympathizes with people who have trouble accepting this concept – especially the part about toy preferences – because it's not something she herself would have believed when she started studying psychology.

That was in the sixties, she recalls. She had begun her undergraduate career as a math major, but in those socially conscious times she was 'looking for more relevance in life' and so she took a psychology course. Psychology provided both relevance and an intellectual challenge for her, and she ended up at the University of California at Berkeley in the 1970s studying individual differences – what makes one person from another. When she started graduate school, she remembers, she believed the environment was responsible for most dissimilarities, but that conviction weakened as she learned more about how mental differences can be related to such biological factors as sex, handedness and 'brain lateralization', or how a person's brain splits up mental duties between its right and left sides. Gradually she became convinced that it would be valuable to study the role biology plays in determining a person's personality and intellectual abilities.

A year after graduating from Berkeley in 1977, while working in a post-doctoral position at the University of Minnesota, Berenbaum stumbled across CAH girls. As part of her studies in behavioral genetics she had been accompanying doctors on their genetics rounds in a Minneapolis hospital. One of the patients was a two-year-old 'boy' that the doctors had just discovered wasn't a male at all. She was a girl who had been exposed to so much testosterone in the womb – because of CAH – that she had been born with male-looking genitals. Her clitoris was the size and shape of a small penis and her labia had fused to form what looked like a scrotum. After discovering her condition, the doctors examined her six-month-old 'brother', only to find that she was really a sister with the same condition. The doctors did surgery on both girls to change their genitals into the female shape, and the girls were put on hormone therapy to stop further effects of the testosterone. With the names changed to reflect their correct sex, they began new lives as females.

Intrigued, Berenbaum talked to a pediatric geneticist at the hospital and asked how many patients they saw like that. 'We get a lot of kids with CAH,' he told her, and that triggered an idea. Here was a nearly ideal way to separate biology from the effects of socialization. CAH females are exposed to exceptionally high levels of male hormones while in the womb – sometimes just as high as male fetuses see – yet they are raised as girls, most of them from birth or shortly afterwards. (Very few of them are as masculinized as the two sisters whom Berenbaum first encountered, and most have ambiguous genitals that can be repaired to look completely female.) Clearly, the male hormones in the womb masculinized the girls' sexual organs to a certain extent, but did the hormones also masculinize their brains? If boys and girls are different only because of the way they are raised, Berenbaum reasoned, then CAH girls should be no different from normal girls. But if some sex differences trace back to the womb, then she should find ways in which CAH girls act more like boys.

It wasn't really a new idea, Berenbaum notes. As early as the 1960s, researchers like John Money and Anke Ehrhardt were studying CAH girls to see if they could spot any lingering effects of the testosterone bath the girls had taken in the womb. In one study, Ehrhardt and Susan Baker interviewed a group of CAH girls and their mothers, asking both mothers and daughters about the daughters' play behavior. These girls had all been detected and treated early, so they had been raised as females almost from birth. Despite that, the CAH girls in Ehrhardt and Baker's study reported play habits very similar to boys. They liked rough, energetic outdoor games and fifty-nine percent of them said they thought of themselves as tomboys. None of the non-CAH sisters of these girls reported themselves as tomboys, however, so the tomboyism didn't seem to be something they learned at home.

None of this was proof that male hormones were responsible for the male-like play behavior, of course. The CAH girls might have been acting like tomboys because of some subtle signals they were getting from their parents, although the parents insisted that they treated their CAH daughters no differently from other girls. Or the CAH girls and their mothers might have been exaggerating the girls' masculine play behavior because they sensed that the interviewers wanted them to answer that way – psychological studies done by retrospective questioning are always open to this type of possible error. But there were other reasons to believe that the CAH girls' unusual hormone environment in the womb was responsible for their masculine behavior. In particular, there were some very convincing studies on monkeys.

Apes and monkeys are man's closest living relatives, and it shows in more than just appearance. At play, juvenile monkeys act much like human children, and one of their most human-like activities is 'rough-and-tumble play'. Two or more monkeys will chase each other around, shrieking, wrestling and pulling on one another's limbs, looking for all the world like a bunch of rowdy elementary school kids at recess. And among the monkeys, as in human children, such rough play is done almost exclusively by the males. The females prefer to play more quietly and let the boys cause all the trouble, in both species.

In the 1970s Robert Goy showed that he could create 'tomboy monkeys' – juvenile females who play like juvenile males – by a laboratory process that mimics what happens to CAH girls. Goy, an animal behavior specialist at the University of Wisconsin, exposed female rhesus monkeys to high levels of male hormones while they were still in the womb. These female monkeys were born with masculinized genitals, much like human CAH girls, and when they got older, they took part in rough-and-tumble play much more often than normal female monkeys. Since no one has suggested that monkey parents raise their male and female children differently, the difference in play behavior must have been caused by male hormone exposure in the womb. More recently, Goy has shown that by applying testosterone at the right time in the pregnancy, he can produce female monkeys that do not have masculinized genitals but that do participate in rough-and-tumble play. Because there is no external sign of these monkeys' masculinization, it's nearly impossible to imagine an environmental explanation.

Goy's monkeys are not a perfect parallel with what happens in humans – CAH girls don't seem to engage in any more play fighting than normal girls even though they are tomboyish in other ways – but the parallel is close enough that it seems unlikely to be a coincidence. The studies have convinced most sex difference researchers that testosterone in the womb creates a propensity for male-like play in childhood: high-energy games, athletic sports, and rough-and-tumble play. So the next time you see a group of kids, especially boys, running around screaming, hitting each other and throwing things, you can tell them they're acting like a bunch of monkeys – juvenile male monkeys, to be exact.

When Sheri Berenbaum first came across CAH females, however, something more cerebral than playtime came to mind. She knew, for instance, that males consistently outscore females on certain spatial tests, such as mental rotations. Would CAH girls and women score more like males or more like normal females on spatial ability? Other researchers, such as Ehrhardt and Baker, had asked that question before but their answers had been inconclusive, either because they had not used tests that focused specifically on spatial ability or because their subjects had been too young. (Some scientists believe the sex difference in spatial ability doesn't become large until after puberty.) So Berenbaum decided that she would give a series of tests, including tests of spatial ability, to as many older CAH females as she could find, and also to their unaffected sisters and female cousins, who would serve as a basis for comparison. In addition she would test a few CAH males who, like CAH females, are exposed to abnormally high levels of testosterone in the womb and who might differ from normal males in some ways. Finally, she would include some brothers and male cousins of the CAH males.

Berenbaum started this study while at the University of Minnesota but didn't have a chance to finish it before she moved to the Chicago Medical School in 1979, so her Minnesota colleague Susan Resnick, who had been working with her on the study almost from the beginning, took it over. After years of tracking down enough CAH patients to make a good study, the two researchers finished in 1982. The results: on tests of general intelligence, they found no differences among the four groups of subjects, males and females, CAH and non-CAH. On spatial tests, as expected, the normal males outscored normal females, and the CAH males scored much the same as their non-CAH male relatives.

But the CAH females showed significantly higher spatial ability than their female relatives, posting scores much closer to the males. It was a startling result, and much harder to accept than the earlier findings on tomboyism. After all, the play behavior can be written off as due to some primitive, unconscious drive – even monkeys do it! – but spatial ability hits much closer to home. How could a shot of testosterone before birth help a person perform better on tests of spatial reasoning twelve or fifteen years later?

One way is that hormones in the womb might modify the wiring of the brain, improving some skills and perhaps worsening others. This is quite possible, Berenbaum says, but she and Resnick wondered if the effect might be less direct. Perhaps, they reasoned, the male hormones alter some brain circuits that influence what a child likes to do. If a CAH girl were more likely to run around from place to place and to pick up objects and manipulate them, she might naturally develop greater spatial ability. To test this idea, the researchers gave all of their subjects a questionnaire asking about which activities they did most in childhood. They found that, like males, the CAH females were on average less interested than normal girls in verbal activities (word games, making up stories and riddles, talking to adults, and so on) and instead preferred things that involved moving objects around. But when Resnick and Berenbaum went past the averages and looked at individual CAH girls – to test, for instance, whether those who had spent more time building models as children had higher scores on spatial ability later in life – they found next to nothing. If the CAH girls were getting their advantage in spatial skills by playing with blocks or erector sets as children, it wasn't obvious.

By now, even though she had moved to Chicago and away from her original subjects, Berenbaum was hooked on the CAH girls. Why were they so different from normal girls, and what other differences might she find? She remembered that the early Ehrhardt–Baker study had also reported that CAH girls preferred stereotypical boys' toys to girls' toys. Many people had doubted those results because the study had been done by questionnaire. Direct observation would be much better, Berenbaum figured, so she designed an experiment in which the subjects would be put in a small enclosed area with a few toys and watched to see which they chose to play with. The whole session could also be videotaped and then later judged by someone who didn't know whether the subject had CAH or not. The only obstacle was that since CAH is so rare – about one in every fourteen thousand births – she would have trouble finding enough two- to eight-year-old subjects around Chicago for an adequate study. That problem was solved when Berenbaum met Melissa Hines, a recent Ph.D. with an interest in sex differences and a job at the University of California, Los Angeles, who was willing to collaborate on the study. Hines would test subjects in Los Angeles and Berenbaum in Chicago, and eventually they would gather data on more than two dozen CAH girls – enough to reach some solid conclusions.

Berenbaum's experimental design was simple. She tested the subjects – CAH girls and boys, and also unaffected siblings and cousins – in their homes. She would set off a small area, perhaps six feet by six feet, with portable screens and then lay out the toys in two rows. The toys were mostly items that other researchers had found to be preferred by one sex or the other. For boys' toys Berenbaum included a helicopter, cars, trucks, blocks and Lincoln logs. 'I'm not comfortable with guns,' she says, 'and a number of parents aren't comfortable with guns,' so the favorite toy of many males was not included. For girls' toys she had dolls, kitchen toys, a telephone, and crayons and paper. The dolls included the popular Barbie, which turned out to be a mistake, Berenbaum says, because 'we found some of the boys were playing with it, but only to undress it.' She also included a few neutral toys – books, a couple of board games, and a jigsaw puzzle.

Under the eyes of an observer and with a video camera running, a child would be put in the enclosure for twelve minutes and allowed to play with whatever toys he or she wanted. On average, Berenbaum and Hines found, the CAH girls spent more than half their time playing with boys' toys and less than a quarter of the time playing with girls' toys. This was almost identical to the boys' performance. but was very different from the toy choices of the non-CAH girls. The CAH girls played with girls' toys only half as much as their normal sisters and cousins did and played with boys's toys twice as much. The researchers found no differences between the CAH boys and their unaffected male relatives.

In a later, follow-up experiment, to guard against the possibility that the children were influenced by the observer's presence to play with the 'right' toys, Berenbaum let each child choose a toy to keep. She gave the child a sack with five toys in it – a car or airplane, a ball, a puzzle, a set of markers, and a doll – and told the child to take the sack to another room and pick out one toy in private. Of course, Berenbaum knew which toy had been chosen by which four were left in the bag, but the child thought it was a secret. 'We've found that about half of the CAH girls choose the transportation toy,' Berenbaum says, and almost none of their non-CAH sisters or female cousins do. That convinces her that the differences she sees in toy selection during the twelve-minute videotaped sessions are real.

What causes the difference between CAH girls and their unaffected female relatives? Berenbaum has tested a number of environmental explanations and doesn't think any of them can account for it. Could the CAH girls, for instance, have been pushed to act more like boys by their parents because the mothers and fathers were aware of their daughters' condition? Berenbaum doubts it. 'It's just as likely that the parents would treat CAH girls more like little girls because they're worried they might grow up too masculine.' Still, she and Hines tested for that possibility, asking the parents such questions as, 'Do you encourage your child to act as a girl should?' There was no difference between the responses of the parents of the CAH girls and those of the parents of the control girls.

Furthermore, if CAH girls were treated differently by their parents because they had masculinized genitals at birth, Berenbaum reasoned that the girls with more masculinized genitals would be more likely to be treated like boys. (One girl was even raised as a boy for the first month, but the rest were identified correctly at birth.) So she and Hines looked for a correlation between the degree of masculinization at birth and toy preference. They found none.

The only explanation seems to be that male hormones in the womb do something to affect a child's toy preferences, just as they lead to tomboyism and higher spatial ability. This result has boggled and bewildered the community of sex difference researchers as no other. The male-like play is understandable and even the spatial ability isn't too hard to swallow, but why should testosterone levels in the womb have anything to do with a child's toy choices? 'That's the sixty-four-thousand-dollar question,' Berenbaum says. 'It's not a 'truck gene', but I don't know what it is.'

The most reasonable explanation seems to be that boys – or girls exposed to testosterone in the womb – are more likely to enjoy manipulating things and watching them move. 'It may have something to do with the reward value of motion,' Berenbaum explains. 'Boys like things that move. They find them more rewarding.' If so, testosterone produces a brain that delights in things that challenge its spatial senses – objects that can be pushed or pulled around, such as cars or airplanes, or that can be stacked in three dimensions, like blocks and Lincoln logs and thus boys naturally gravitate to such toys, while they hold no special interest for most girls. Yet another part of the answer, Berenbaum speculates, may be that because boys have better gross motor skills and girls are more adept at fine movements, the sexes may choose toys that reflect this – boys prefer toys that allow them to make big movements, like flying a jet through the air, while girls pick activities that demand fine control, such as drawing a picture or dressing a doll. (Although Berenbaum didn't study whether CAH girls like guns, there is at least one obvious reason that guns might naturally appeal to boys more than girls: playing with guns involves aiming at a target, which would exercise one aspect of a child's spatial skill.)

Over the next few years Berenbaum expects to learn much more about CAH girls because several states, such as Texas, have begun to keep registries of all CAH births. This will give her a much larger group of children to study in her quest to pin down testosterone's effects on the brain. However, the general picture is already clear, she says. In CAH girls, and presumably in males as well, testosterone in the womb masculinizes several mental and psychological traits – play activities, spatial ability and even the choice of toys.

Berenbaum is just one of many researchers who follow a similar line in sex differences. Since it is the presence or absence of male hormones in the womb that determines whether a fetus will turn down a male or female path, these scientists study what happens when someone is exposed to an abnormal hormone mixture during development. In practice this usually means looking at females who received higher-than-usual levels of male hormones, such as CAH girls, or males who got lower-than-normal amounts. This is often a valuable approach, as Berenbaum's work proves, but it has its limitations. Because it is unethical to experiment on human fetuses, researchers must depend on those cases where an accident or defect caused the fetus to be exposed to an unusual hormone mix – so-called 'experiments of nature'. Not only are such cases quite rare but researchers have no way to control the hormones in the womb or know exactly what the hormone exposure was.

For these reasons, many sex difference researchers rely instead on laboratory animals, especially rats, mice or guinea pigs. Although monkeys are more closely related to humans, they are expensive to buy and care for, take a long time to mature and have complicated behavior. In contrast, rodents are cheap, easy to work with and show clear, objective differences between the sexes (remember Gorski's rat sex movie). Rats are particularly valuable because their brains continue to mature after birth, so researchers can avoid fiddling with the hormone mix in the womb and simply manipulate the hormones after the rats are born. This entails castrating newborn male rats and removing the ovaries from ('ovariectomizing') newborn females in order to stop the normal production of sex hormones, and then injecting the rats with the desired doses of testosterone, estrogen or whatever. The researcher, not the rat, determines what hormones the rat's brain is exposed to during the first few days of life.

Using this technique, Christina Williams at Barnard College in New York City has proved a connection between the way rats master a maze and what sex hormones their brains are exposed to during development. In a series of experiments that her peers praise as 'elegant' and 'convincing' (the highest compliments in the scientific community), Williams showed that rats given testosterone shortly after birth depend more on geometric cues and less on landmarks to learn their way around a maze than rats not exposed to the male hormone. Combined with the work of Thomas Bever on differences in the way men and women find their way around a maze (described in Chapter 1), Williams's results imply that it may be hormone mixtures in the womb that turn men toward maps and women toward landmarks.

Like Sheri Berenbaum, Williams is one of the band of young female scientists who have been drawn, for one reason or another, into the study of sex differences over the past decade or so. She's always been a bit of a rebel, she says. 'I grew up protesting the [Vietnam] War, living outside of Harvard, wearing beads.'

In her senior year of high school, she wanted to go to college at Cornell because 'in those days it was thought of as a hippie place'. By chance, though, she heard of a small college in western Massachusetts called Williams College. Traditionally all-male, it was planning to let women in the next year for the first time, and the thought of being one of the first women to infiltrate a male bastion caught her fancy. 'I always wanted to be different.' That, plus the fact that she and the college shared the same name, convinced her to visit the campus that fall when the leaves were turning, just to see what it was like. 'I thought it was the perfect set for a college. I was a rather dramatic young lady and I thought the backdrop was important.' In the fall of 1971, she was among the one hundred and fifty women who joined about twelve hundred men for classes there.

Five foot two, blonde and attractive, she sometimes found it hard to be taken seriously by members of the still predominantly male faculty, who didn't know quite how to deal with the female influx. An English professor told her that she shouldn't decide to be an English major 'just because you think that's what girls should do.' A chemistry professor, wanting to keep that major mostly male, said, 'I hope you're not planning to take any more chemistry classes.' But other professors were more supportive, she remembers, and she found a home in psychology when she took a course from Tom McGill, an animal behavior specialist who had done work on sex differences in mice. As part of that course she performed an experiment evaluating the sexual behavior of male mice with brain lesions. She would put female mice into the cage with the males and watch them have sex. 'I thought it was great. It was especially titillating to call home and tell my parents what I was doing at college.'

That titillation matured into a long-term affair. Majoring in both biology and psychology, she worked for McGill as a teaching assistant for the next three school years and served as a research assistant in his lab for a couple of summers. Her only break came during the summer after her junior year, when McGill pushed her to get a job working for someone else to get a taste of other areas of psychology. She spent the three months working with emotionally disturbed children. 'I hated it,' she remembers. Just when she thought a child was beginning to improve, it would be time to send him back home – where all of the good they had done would probably be undone. 'I found that I wasn't a good clinician,' she says. 'I wanted to fix their brains, to change the world, not send these kids back home to troubled families.'

From Williams College she went to the Institute of Animal Behavior at Rutgers University in Newark, New Jersey, for a Ph.D. and then spent a year at Johns Hopkins in Baltimore before ending up at Barnard College in 1981. In graduate school Williams found that she loved to work with baby rats and study their development, and much of her early research focused on feeding behavior and its neurochemical controls, such as how a newborn rat's instinctual suckling behavior is triggered. From there she moved into female sexual behavior (such as the back-arching lordosis response) and sex differences, but she didn't begin the research that would gain her the most attention until 1985. It was then she met Warren Meck, a psychologist who had recently moved from Brown University to Columbia, next door to Barnard.

'I wasn't sure whether I hated him or liked him,' she remembers, but she was interested enough to agree to go out to eat with him. The meal was a success, both personally and professionally. Not only did it begin a friendship that would lead to marriage three years later, but in looking for projects that would combine their talents and interests, Williams and Meck mapped out two lines of research that would keep them busy for years to come. One line dealt with sex differences in the effects of the chemical choline on memory in rats. Meck has been the spokesman for the team on that research. The other venture was to explore how sex hormones affect maze learning in male and female rats. Williams is more closely identified with this work, but she insists it's an equal collaboration. 'I think of the work as *ours*.'

The maze that Williams and Meck use in their work is a twelve-arm radial maze. This is not the type of maze you see in cartoons, where the rat must find its way along a confusing zigzag path littered with wrong turns and dead ends. Instead, picture a small, round pedestal table about three feet high and eighteen inches across, with a dozen equally spaced arms pointing straight out from it, each of them about thirty inches long. A rat is placed on the round center and must walk out to the end of an arm to reach the bit of food placed in a shallow well, and then walk back to the middle before heading out along another arm. Williams and Meck put food at the end of only eight of the arms – always the same eight arms – and then train the rats to find all eight food pellets in as few trips down the arms as possible. The rats are put in the maze once each day until they reach a steady state – until their performance stops getting better with practice. (On average, peak performance means the rat needs about ten to twelve trips to get all eight pellets, since even a well-trained rat will sometimes try an unbaited arm or forget where it has been and go down an arm it has already visited.)

Williams and Meck found their first sex difference in how quickly male and female rats learn the maze. Although the sexes do not differ in how well they eventually master it – both sexes average the same dozen or so arms to find all eight baits – males reached steady state a few days earlier than females. What was causing the sex difference? If the male and female rats were using different strategies to learn the maze, then even at steady state, when they navigated the maze with equally few errors, the two sexes were still doing something different. The trick was to uncover it. 'We saw the sex differences disappear at steady state, and we wondered how we could get them back,' Williams says.

They remembered the work of other rat researchers, who had found that rats orient themselves inside a maze by using cues from outside – the shape of the room, or the tables and other fixed objects inside the room – and they thought that their rats were probably doing the same thing. Many times a rat in the middle of the maze would look around the room, apparently at the furniture or perhaps at the walls, before it would decide which arm to go down next. Or the rat might creep to the edge of the circular table and peer down at the floor. Sometimes, Williams says, a rat that had gotten the food from seven of the arms would take fifteen seconds or more in the center of the maze, looking around and thinking, apparently trying to remember which arm held the last bait. 'You think it's never going to remember,' she says, and then all of a sudden it heads down the correct arm.

What would happen if some of the external cues were removed? The rats' performance in the maze should get worse, but would it affect the males and females differently? Williams and Meck decided to find out. To get rid of the 'geometric' cues, they jury-rigged a circular curtain that could be put around the radial maze. This would hide the corners and walls of the rooms so that the rats couldn't use them as compass points to navigate the maze. To eliminate the 'landmarks' – a table with a computer, a cart with rat cages, some chairs – they would either move them around to new positions or remove them from the room entirely. Each day the rats were tested with a different set of geometric cues and landmarks. Sometimes the room was as usual. Other times the furniture was moved around, or the furniture was in its usual places but the curtain was up. On still other days, both the curtain was up and the furniture was shifted. And in each setting, the rats would be put in the middle of the maze and watched as they tried to find the food.

When Williams and Meck added up the results, a striking difference between the sexes emerged. Getting rid of the landmarks didn't hurt the male rats' performance, but putting the curtain up did. Unless they could orient themselves by the room's walls, they had much more trouble remembering where all the food was. The female rats, on the other hand, did fine as long as they had either the room geometry or the normal landmarks to work with. Their scores dropped significantly only when both the curtain was up and the furniture had been moved around.

The male and female rats were using very different strategies to learn the maze, the researchers concluded. The males were single-minded, focusing only on a global view in which the room provided a north-south-east-west orientation. The females seemed to rely on both geometry and landmarks equally and they did poorly only when they no longer had either set of cues to work with.

'It isn't that males can do something and females can't, or that females can do something and males can't,' Williams says. Both sexes learn the maze equally well – they just do it in different ways. In a sense, the males take the 'quick and dirty' route to learning the maze, not doing any more work than they have to. This may explain why the males take fewer training sessions to reach peak performance, Williams says, but it makes them more vulnerable to removal of cues. In contrast, the females are slower but more thorough. They use more information to learn the maze, and although they may take longer to reach the same peak performance, it pays off when they are forced to run the maze with some of the cues gone.

Having nailed down the sex difference in maze performance, Williams and Meck set out to see if it could be traced back to differences in the hormones that the developing rat brain is exposed to. They castrated male rats to deprive them of testosterone and injected females with a sex hormone to masculinize their brains, then performed the same maze-learning experiment with these cross-sex rats. The result was what one might guess from Berenbaum's work with CAH girls: the hormone-treated females performed like males, and the castrated males acted like females. Now it was the females, whose brains had been masculinized, that learned the maze with fewer errors but couldn't cope when the circular curtain was put in place, and it was the males, whose castration had demasculinized them, that took longer to reach the same level of accuracy but paid attention to both landmarks and geometric cues.

To determine which particular area of the brain is responsible for the maze learning, Williams castrated male rats and implanted small hormone capsules into specific sections of their brains. When the implants were put into the hypothalamus, a part of the brain involved in emotion and instinctive behavior, they had little effect. But when they were placed in the frontal cortex or the hippocampus, two areas of the brain that control learning and memory, the castrated male rats performed in the maze just like normal male rats. The hormone, Williams concluded, must be directing the growth of neurons in those particular parts of the brain.

This series of experiments by Williams and Meck offers the most complete picture yet of a cognitive difference between males and females. Rats use different strategies to learn a maze depending on their sex, with male rats relying mostly on geometric cues and females using both landmarks and geometry to learn the maze, and the difference is produced during the first few days after birth by sex hormones acting on the hippocampus and frontal cortex. All that remains to finish the picture is to pin down exactly how the hormones change the brain – something that other researchers are working on, but which is still cloudy at this time.

Although Williams and Meck didn't know it at the time as they were tracking down this sex difference in maze learning in rats, Thomas Bever at the University of Rochester was on the trail of an amazingly similar difference in humans. As described in Chapter 1, he trained male and female college students to find their way around a computer maze and then watched what happened when he modified the maze by removing the landmarks or altering its geometry. Without the landmarks, the women did much more poorly, while the men's performance stayed the same. The reverse was true when Bever lengthened some of corridors in the maze – the women hardly noticed but the men's times slowed down.

Bever performed equivalent experiments on rats (with a real maze, not a computer maze) and found a very similar sex difference, with the males depending on geometric cues and the females on landmarks. Although he hasn't done the same hormone manipulations on the rats that Williams and Meck did, it's a safe guess that male hormones acting on the rat brain are at the heart of this sex difference, too.

It could be just a coincidence that human males and females show the same difference in finding their way around a maze that male and female rats do, but Bever and Williams don't think so. Although it's always dangerous to generalize from rats to humans, both of them suspect that the levels of testosterone in the womb may go a long way toward explaining why men are happier using a map to find their way around a new place while women are more likely to orient themselves with buildings, signs and other landmarks.

Much of the work done on rats and other lab animals has focused, like Williams and Meck's work, on how male hormones masculinize the brain during development. Just as it does in monkeys, for instance, testosterone has been shown to produce a rough play behavior in juvenile rats, and other researchers have found that male hormones cause rats to be more aggressive as adults. There is much more of this research on testosterone's influence – too much for us to do more than wave at it as we go past – but the overall pattern that emerges from it is consistent with what we've already seen: exposing the developing brain to male hormones causes the animal to behave in certain 'masculine' ways characteristic of its species. In general, in a non-human mammalian species, if the male has a behavior pattern different from the female, it's a good guess that the difference can be traced back to the level of male hormones during development.

Having said that, however, it's important to note that female hormones also play some role in influencing behavior. Although that role is not nearly as well understood as testosterone's, scientists are finding that estrogen does seem to feminize the brain in certain ways.

One of the best studies of estrogen's influence was done on rats by Jane Stewart of Concordia University in Montreal. When placed in an unfamiliar open area, female rats are much braver in exploring it than males, moving from place to place while the males huddle in a corner and peer about. (Although this may seem to contradict the stereotype of the male being more aggressive, the willingness to venture out into a new area is actually quite distinct from aggressiveness, and it is truly a 'feminine' trait in rats.) Stewart discovered that this feminine open-field behavior depends on exposure to estrogen in the first few weeks of life. If she removed a female rat's ovaries in the first eight days after birth, the female rat was just as timid as a male in exploring a new area, but if she gave the ovariectomized rat estradiol (an estrogen-like hormone) between days ten and forty after birth, it boosted her courage. Stewart's conclusion: the female rat's greater willingness to explore a novel environment is due not simply to the absence of testosterone during development but depends upon the estrogen produced by the rat's ovaries.

In tracing how estrogen affected the rat's brain, Stewart discovered one detail that at first sight doesn't appear to fit in. If a young, ovariectomized rat is given too much estradiol, it doesn't make her more 'feminine' – that is, braver in the open field – but instead less adventurous. Higher doses of estrogen actually make the rats less feminine (or more masculine, depending on your point of view).

What's going on? The apparent paradox arises because – as we saw in the last chapter – the actions of sex hormones in the brain aren't as simple as they might first appear. When testosterone enters the brain, some of it is transformed to estrogen and much of the masculinizing influence of testosterone is actually caused by estrogen that has been converted from testosterone. In rats, for example, although rough play in juveniles is caused by testosterone acting directly on the brain, male sexual behavior is produced by testosterone that has first been converted into estrogen. And Christina Williams found that she could produce male-like maze learning in rats by treating them with estradiol instead of testosterone. (Given that it is estrogen compounds which are responsible for much of the masculinization of the brain, some researchers like to avoid the distinction between 'male' and 'female' hormones altogether. Male bodies may produce more of one type and female bodies more of another, but both types of hormones are used by both sexes.)

This situation can make it difficult to say exactly where masculinization stops and feminization begins. In Stewart's experiment, the rats apparently needed a certain amount of estrogen to act like females – to move around confidently in an open area – but if they got too much estrogen it masculinized them and caused them to be more cautious. Although researchers are still debating the issue, some suggest that this may be a general pattern – that a certain minimum amount of estrogen is needed to make a brain female, and that higher levels of estrogen will actually masculinize a brain in many ways.

Despite the time and expense needed to work with primates, a few researchers have looked at the influence of sex hormones on monkeys' brains, and their findings reinforce the picture drawn by the rat work. For instance, at the National Institute of Mental Health in Bethesda, Maryland, Jocelyne Bachevalier and Corinne Hagger found certain learning differences between male and female infant rhesus macaque monkeys and were able to trace them to the presence or absence of male hormones in their brains. They tested the macaques on 'object discrimination', in which the monkeys had to learn which of a pair of objects hid a bit of food. At three months of age, the researchers found, female monkeys learned to do it much faster than male monkeys, but the difference had disappeared by six months. Apparently the part of the brain that controls this ability – the inferior temporal cortex – matured earlier in females than in males. When male monkeys were castrated at birth, they performed just like normal females. When female monkeys had their ovaries removed at birth and were injected with dihydrotestosterone, a hormone closely related to testosterone, they did no better than normal males. Thus this male hormone seems to slow down the development of at least one part of the brain in monkeys.

There may be something similar going on in humans, Bachevalier and Hagger suggest. In general, girls develop a number of abilities faster than boys and there is some evidence that the female brain matures earlier. Parents will often tell you, for instance, that girls seem to talk earlier than boys. Even after birth, sex hormones may still be directing the primate brain down either the path marked 'male' or the one marked 'female'.

Of course no matter how much evidence is accumulated about rats or monkeys, it only suggests possible explanations for sex differences in humans. It doesn't prove anything. To understand the differences between human males and females, one must eventually turn to men and women, boys and girls. That's why Berenbaum's and others' work on the CAH girls is so important, and that is why researchers have scoured hospitals and medical journals, looking for examples of other people hit with unusual concentrations of sex hormones while their brains were developing. Besides the CAH females, doctors and scientists have studied several other such syndromes and each offers its own insights into how hormones steer the brain.

At the Cornell University Medical College in New York City, Julianne Imperato-McGinley investigates the victims of two feminizing syndromes that affect genetic males. Comparing the two groups gives her a chance to identify which particular hormones participate in the development of such mental traits as spatial ability.

Imperato-McGinley first came to the attention of the sex difference world in 1979 when she described a group of interrelated males living in a small Caribbean country who suffered from 5α-reductase deficiency, the very rare condition described in Chapter 3 that causes males to look like females at birth but to develop a penis and testicles at puberty. More recently, she has returned to the same Caribbean country to study a group of interrelated XY women, all from the same small town. Like Maria Patiño, the Spanish hurdler barred from athletic competition because of her chromosomes, these women are genetically male but thanks to a genetic defect their bodies cannot detect male hormones and so they develop as women, albeit without ovaries or uterus. In terms of hormone exposure, the XY women offer a perfect counterpoint to CAH women, since CAH females are subjected to exceptionally high levels of testosterone in the womb while XY females see absolutely no testosterone at all.

Imperato-McGinley and colleagues have tested a group of XY women on a Spanish version of the Wechsler Intelligence Scale, along with male and female relatives of the women to act as controls. Since XY females look completely female at birth, even though they are genetic males, they are raised as girls and should have the same socialization as girls. Thus, if socialization explains differences in test scores between males and females, the XY women should score like normal women, including having lower spatial ability than men. And indeed, Imperato-McGinley found that the XY women do score worse than the men on tests of spatial intelligence – but they also scored worse than their normal female relatives. As far as spatial ability goes, they are 'super-feminine'.

What's going on? The apparent explanation, Imperato-McGinley says, is that male hormones in the womb do improve a person's spatial ability, and even a small amount has an effect. Although normal females are not exposed to nearly as much male hormone during development as males, they still do get a low dose of the hormones, and that seems to be enough to give them an advantage over those who, in practical terms, get absolutely no male hormones at all: the XY women.

At the clinic where Imperato-McGinley and her colleagues work with the XY females and the 5α-reductase deficient males, a running joke is that there's no need to give a special chromosome test to identify the XY women – just hand them a jigsaw puzzle. The clinic keeps jigsaw puzzles in its waiting area to entertain the patients, and the males with 5α-reductase deficiency love them, Imperato-McGinley says. They'll happily attack a two-thousand-piece puzzle and finish it off quickly. But the XY women have trouble piecing together even a hundred-piece puzzle and complain that it gives them a headache. Since working a jigsaw puzzle is a rough test of spatial ability – it demands visualizing how one piece fits into another – this difference between the two types of patients provides a graphic demonstration of how male hormones influence spatial skills.

Following up on the clue from the jigsaw puzzles, Imperato-McGinley has tested spatial ability in the males with 5α-reductase deficiency, and her preliminary results indicate that their spatial skills are at least average for males and perhaps above average. This difference between them and the XY women allows Imperato-McGinley to deduce exactly which male hormone is responsible for developing spatial ability in the womb.

XY women's bodies do not respond to any male hormones, either testosterone or dihydrotestosterone, while the 5α-reductase deficient males are missing only dihydrotestosterone. Since the latter group's spatial ability is fine, dihydrotestosterone cannot be essential for developing spatial skills. This implies, Imperato-McGinley says, that it is probably testosterone, the other major male hormone, which holds the key to working jigsaw puzzles and doing mental rotations in three dimensions.

Another clue to the hormone roots of spatial ability comes from Turner's women, who are born with only a single sex chromosome, an X chromosome, instead of a pair. They are usually short and may have a mild physical deformation, such as 'webbing of the neck' (loose folds of skin on the neck), but otherwise they appear normal, and they do just as well as normal XX women on tests of verbal ability. But don't ask them to find their way around a city with a map. They score much more poorly than other women on most tests of spatial ability, and their problems in getting lost are legend among sex researchers. One Turner's woman consistently got lost going from the elevator to her doctor's office even though she had visited him time after time.

Like XY women, Turner's women's brains are exposed to much lower levels of male hormones than normal females, but for a different reason. With only one X chromosome, their gonads fail to develop into either testes or ovaries, and their bodies get only the minor amounts of sex hormones produced by other glands. Once again: minimal male hormones, minimum spatial ability.

Turner's women also offer an insight into the function of female hormones. Besides their problem with spatial skills, Turner's women have weaknesses in short-term and long-term memory, word fluency, and recognizing emotions in the faces of others – all things in which women normally excel – and the reason may be a deficit of female hormones since Turner's women, without functioning ovaries, are not exposed to the normal amount of estrogen during development and later in life. Just as Jane Stewart found that female rats need a certain amount of estrogen to develop their normal open-field behavior, it's possible that the brains of human females must be exposed to a minimum of female hormones in order to fully develop such abilities as memory and word fluency. There is, however, not yet enough evidence for scientists to agree on this idea.

In addition to such 'experiments of nature' in which something goes wrong in the normal process of making a person male or female, there is another situation that exposes a fetus to unusual sex hormones, but it's one that few people would call abnormal: opposite-sex twins. When both a male and a female fetus occupy the womb, the female twin may be exposed to some of the testosterone produced by the male. Could this have some effect on the twin girl?

A couple of studies suggest that it might. Susan Resnick, Sheri Berenbaum's original partner in studying CAH girls, has looked at sensation seeking in opposite-sex pairs of twins. Sensation seekers are those who look for new and exciting, sometimes scary experiences, such as trying new foods, water-skiing, or jumping out of a plane with only a thin sheet and a few ropes keeping you from crashing into the ground at hundreds of feet per second. Males consistently score higher on tests of sensation seeking than females, and when Resnick tested several hundred female twins on sensation seeking, she found that those with twin brothers scored significantly higher than the ones with twin sisters.

And at the University of Colorado, researchers examined spatial ability in women who were twins. They found that on a mental rotations test, females with male twins scored much higher than did a control group of women with non-identical female twins. Indeed, after a little practice, the females with male twins scored just as high on the test as did their twin brothers.

CAH girls, XY females, Turner's women and females with male twins – in each we seem to be hearing echoes of the womb, as their hormone exposures make themselves felt years later. But isn't it possible that the echoes are from a nearer source, the girls' upbringing and socialization? In opposite-sex twins, for instance, perhaps being raised with a twin brother causes a girl to do more of the things that develop spatial ability or teaches her to be more willing to try new things. And maybe CAH girls equal males at spatial ability because their parents, having seen the masculinized genitals at birth, treat the girls more like boys. It's harder to explain why XY females and Turner's females should have poorer spatial skills than normal females, or why Turner's women have weaknesses in memory and recognizing facial expressions, but with a little imagination even that is possible. You might hypothesize, for example, that XY women try to compensate for their lack of a monthly menstrual cycle by being super-feminine.

Is there a way to choose between the possible explanations? Yes. Scientists have several rules of thumb that work pretty well in determining which of several competing theories is most likely to be correct. One is referred to as 'Occam's razor', and it simply means that, when everything else is equal, you should choose the simplest, most concise explanation, the one that can explain the most things with the fewest rules.

Consider the facts that CAH girls play like boys and juvenile female monkeys exposed to testosterone in the womb play like juvenile male monkeys. You can explain the monkey case with hormones and the human case with socialization, or you can suppose that both arise from the same underlying biological principle. Or look at the parallel between how Christina Williams's male and female rats learn their mazes and how Thomas Bever's undergraduates learn theirs. Again, you could suppose it's just coincidence, or you could hypothesize that the same thing is at work in both cases – something about the male brain (in humans and in many species of rodents) likes to navigate with some sort of north-south-east-west system. And on and on. Many of the sex differences we see in humans, from play behavior and spatial ability to nurturance and aggressiveness, have parallels in other animals. The rule of Occam's razor suggests that human sex differences arise, at least in part, from the same source that creates the sex differences in animals – hormones in the womb.

Another rule of thumb that scientists use to choose among competing theories is to pick the one that predicts the most things. The more predictions a theory makes, the more chances it has of being proved wrong, and thus the better tested it will be. The socialization theory doesn't really make predictions about the 'experiments of nature' – it offers explanations. Given that society treats boys and girls differently, what would you predict about how parents will treat CAH girls? You might guess that the parents would treat the CAH girls more like boys, or you might guess that they'd push them harder to act like girls, but both positions would be no more than guesses. And why would the socialization theory predict XY girls – who look just like normal females and who usually aren't diagnosed until puberty – to be treated any differently from normal girls? In each case, if you know the answer – CAH girls act more like boys, XY women are 'super-feminine' in their spatial ability – then you can work backwards to offer some explanation of how socialization might explain it, but you haven't really predicted anything.

The biological explanation seems much more compelling to many scientists because it makes clear, testable predictions. Exposure to high levels of male hormones in the womb will masculinize a person, male or female, and conversely, if a fetus is exposed to low levels of male hormones, the person should not have masculine traits. Thus, if CAH girls behaved in a more feminine manner than other girls or if Turner's women or XY women were more masculine, this would be a flat contradiction of the hormone hypothesis. Given a choice between a theory that makes no predictions unless you know the answers ahead of time and one that makes testable, correct predictions, scientists tend to choose the latter.

For these reasons, more and more researchers are accepting the idea that sex hormones play a role in influencing behavior later in life. Furthermore, there is one other 'experiment of nature' that seems to have no good environmental explanation at all.

Idiopathic hypogonadotropic hypogonadism is a rare syndrome in males that suppresses puberty. Males with IHH don't produce enough of the hormones that stimulate the gonads ('gonadotropins'). The boys develop normally in the womb, apparently because their mothers provide enough gonadotropins to stimulate their fetal gonads and masculinize their sex organs, but after birth their testes are inactive. The males look normal at birth, have no obvious physical or mental defects, and grow like normal boys – and thus are exposed to the same environmental influences that other boys are – but because their testes don't produce testosterone, puberty never starts. Once they have been diagnosed, they are treated with hormone-replacement therapy to artificially induce puberty.

In 1982, Daniel Hier and William Crowley reported that they had tested nineteen IHH males against nineteen normal controls. The two groups did not differ on tests of verbal ability, but on the three spatial tests Hier and Crowley used, the contrast was striking. On the Block Design subtest of the Wechsler Adult Intelligence test, control males got an average of thirty percent more right answers. On the Space Relations subtest from the Differential Aptitude test, control males were up by more than forty percent. And on an embedded figures test, where the subjects were timed on how long they spent getting an answer, controls were nearly twice as fast as the IHH males.

The two doctors also found that the patients with the more severe cases of IHH – the ones whose bodies produced the least male hormones – were the ones with the worst spatial ability. The size of a patient's testicles is a good indication of how severe the disorder is, since the less gonadotropins the body produces, the smaller the testicles. By measuring the volume of their subjects' testicles, Hier and Crowley showed that on average, the patients with the smallest testicles had the poorest spatial ability.

To determine when in life the male hormones exerted their influence, Hier and Crowley looked at five patients with acquired hypogonadotropic hypogonadism, or AHH, a disorder similar to IHH, but one that starts after injury or illness. All of these patients had contracted AHH during or after puberty, so they had been exposed to normal levels of male hormones at least until puberty. Given the same tests as the others, these five got scores that were almost identical with the normal males, implying that their spatial ability had not been harmed by losing the function of the testes after puberty. The conclusion, Hier and Crowley say: Testosterone or other male hormones produced by the testes are responsible for developing spatial ability, and they do their job sometime between conception and puberty.

In addition to these various 'experiments of nature', researchers have also looked for clues to sex differences in what one might call 'experiments of medicine'. From the 1950s to the 1970s, doctors in the United States regularly prescribed synthetic hormones to pregnant women who had a high risk of miscarriage. The practice stopped when doctors discovered the treatments had some rare but nasty side effects, including birth defects and increased risk for certain types of cancer in the child. The best known of the synthetic hormones was diethylstilbestrol, or DES, a synthetic estrogen, but there were many others. Some of these chemicals were natural and synthetic versions of progesterone, a hormone involved in regulating a woman's reproductive system during pregnancy, and others were chemically related to androgens, or male hormones. Thus the chemicals affected the body in various ways, some producing effects similar to those of female hormones and others more like male hormones. Since there were many thousands of babies born who had been exposed to these synthetic hormones in the womb, they offer the largest hormone experiment on humans ever done – and ever likely to be done.

June Machover Reinisch, former director of the famous Kinsey Institute for Research in Sex, Gender and Reproduction, has been researching these subjects for nearly twenty years. The evidence is mixed and sometimes contradictory, she notes, mostly because there was no standard treatment for the women who took these synthetic hormones. Different doctors used different synthetic hormones, often in combinations of two or more, and they varied the dosage, the duration and the point in the pregnancy at which the hormones were given. Still,

Reinisch says, a consistent pattern has emerged after two decades and dozens of studies.

Few patients were given androgen-like hormones with no other hormones to confuse the issue, but those few cases are telling. Male children exposed to high levels of these male hormone-like chemicals were much more physically aggressive than brothers who had not been exposed to these synthetic hormones in the womb. Girls exposed to the same hormones in the womb were more masculine in a number of ways. They liked rough play, boys' toys, wearing pants instead of dresses and playing with boys instead of other girls – in a word, they were tomboys. When asked whether a career or marriage was more important, they were more likely than other girls to pick a career. Even when mixed with a progesterone-like hormone, which has complicated effects on the fetus, the androgen-like hormones still made both boys and girls more masculine, and especially more physically aggressive.

The effects of estrogen-like hormones, particularly DES, are more complex. DES boys seem to be somewhat less masculine and more feminine than other boys – they appear to be less interested in contact sports, have lower spatial ability, and are more likely to express interest in writing or social services. Somewhat paradoxically, if DES does anything to female children, it may sometimes masculinize or defeminize them, especially their sexuality. There is some evidence that DES women may be more likely to be bisexual or homosexual, although it's still very tentative.

DES and other synthetic estrogens were likely to be mixed with other hormones, complicating and perhaps concealing their effects, but Reinisch says she still found a consistent – though weak – pattern in children exposed to these hormone mixtures in the womb. Both males and females were feminized or demasculinized in some ways. The boys were less athletic and less assertive. The girls didn't like rough play, were more interested than other females in marriage and children, and in general were more 'girly'.

Putting it all together, Reinisch says that exposure in the womb to sex hormone-like chemicals does affect a child's behavior and intellectual abilities in certain ways, sometimes making the child more masculine, sometimes more feminine, depending on the chemical. The study of women given synthetic hormones during pregnancy is not a perfect experiment by any means, but it's likely to be the most extensive experiment on humans we'll ever have.

* * *

The evidence is overwhelming: testosterone, estrogen and other sex hormones do something to the brain before and after birth to change it, to push it in either a 'masculine' or 'feminine' direction. Experiments show this explicitly for rodents and primates, but the evidence is nearly as solid for humans. Data pile up on data – CAH girls, XY females, Turner's women, IHH men, DES babies, and so on – and more than anything else, it is the consistency of the data that is most convincing.

Study after study indicates that exposure to testosterone in the womb increases spatial ability later in life and that a testosterone deficit leads to poor spatial ability. Male hormones seem to play a role in spatial skills as diverse as orienting oneself with a map, mentally rotating three-dimensional objects and working jigsaw puzzles, and because of the importance of spatial ability to higher mathematics, male hormones may also contribute to the male advantage in math, although that is still highly controversial.

Nearly as well established as testosterone's role in spatial ability is its connection with rough play in youngsters, and, in general, male hormones seem to be behind much of what we think of as 'boyish' behavior in children, from young males playing with trucks and building blocks to little girls being tomboys. Work with lab animals shows that fetal testosterone can increase aggressiveness later in life, and studies on children whose mothers took synthetic hormones indicate that this may be true for humans as well.

The effects of estrogen are not nearly so agreed upon, since less work has been done on them, but female hormones may be necessary to fully develop a number of traits, including word fluency, memory and recognition of emotions in faces.

There are various details and complications that are not yet understood. In particular, since testosterone is converted into estrogen in some parts of the brain, there is a fine line between masculinization and feminization. Some researchers believe that – for some traits at least – estrogen and testosterone can do similar jobs, and the difference between masculinization and feminization comes down to how much hormone exposure the brain gets. Low levels point the brain down the female path and high levels push it on to the male route. And the work of Jane Stewart and others indicates that there may be optimal levels of hormones for various traits and abilities – either too little or too much in the womb and performance later in life can be impaired.

Still, the general pattern is clear. Sex hormones influence the developing brain, shaping various mental and psychological traits as well as the physical body. That influence may be modified by events later in life, but years later echoes of it can still be heard.

CHAPTER 5

My Brain's Bigger Than Your Brain

Are men's and women's brains different? The very question is enough to start food fights in faculty lunch rooms at major universities across the country and even (as I've discovered) disturb the harmony of an otherwise peaceful meal at home with your spouse. In other words, it's a sensitive subject, as many of the scientists who do research in the area can testify. But we're safe here – no bread, no salad, no spaghetti, nothing to throw but this book. So let's talk about it.

Sex hormones shape the body, forming a male or female reproductive system in the womb and making further changes at puberty. That much is undisputed. And as we saw in the last chapter, sex hormones in the womb also influence mental and psychological traits later in life – childhood play behavior, spatial ability, verbal skills and so on. This is not exactly undisputed (some hold-outs still insist that it's possible to explain everything in terms of socialization), yet as the evidence continues to pile up, fewer and fewer scientists are doubting that many of these sex differences have their origins in the womb. But if so, then the hormones in the womb must be creating some physical differences between male and female brains – differences that make their presence felt years later when a little girl chooses between a doll and a truck or when a little boy decides whether to sit quietly and look at a book or run around the room bouncing off the walls.

And if you stop to think about it, there's really no other way it could happen. If, for instance, the extra testosterone that CAH girls get in the womb really does lead to tomboyism and improved spatial ability later in life, then the testosterone must be causing some changes to their brains that make them more masculine. Yet for some reason the thought of real, physical brain differences between the sexes seems to bother people much more than the equivalent idea that men and women are different and that hormones in the womb are somehow responsible. The existence of tangible dissimilarities in the brain seems more personal, more serious, more threatening than mere differences on psychological tests.

Because of that, the search for sex differences in the brain has taken on a significance much greater than its simple scientific import. The question of whether there are detectable, measurable differences between the male and female brain has become in some ways that ultimate indicator of the reality of sex differences.

So it's not surprising that the longest running, most contentious dispute in modern sex difference research has centered on claims that one particular part of the brain is different in men and women. The debate over the corpus callosum, the band of tissue that connects the two hemispheres of the brain, began more than a decade ago and is still going strong.

In the late 1970s, Marie-Christine de Lacoste was a graduate student in physical anthropology at Columbia University in New York City doing research on the evolution of the human brain. She had never intended sex differences to be part of her dissertation. Plans changed, however, when she moved to Dallas (for 'personal reasons') before finishing her Ph.D. In Dallas she took a job as a research assistant at the University of Texas Southwestern Medical Center, but she still wanted to complete her doctorate so she looked around for a dissertation topic she could work on in her new location.

Trained to look at the human brain, de Lacoste decided to study how damage to various parts of the brain affects the corpus callosum. Most of the communication between the right and left halves of the brain goes through the callosum, and if a section of the brain is destroyed by a stroke, for instance, then the part of the corpus callosum carrying signals from that section should degenerate. By studying the pattern of degeneration, de Lacoste hoped to learn more about which parts of the corpus callosum carry signals from what parts of the brain.

So de Lacoste began collecting autopsied brains, both from people who had suffered brain damage and from healthy-brained controls. Then one day serendipity struck. She had assembled fourteen undamaged specimens – nine male and five female – and she was lying on her bed, looking over pictures of those brains, neatly dissected and displayed, when 'I just happened to notice that there were very different corpora callosa in the men and the women.' (Corpora callosa is the Latin plural of corpus callosum.) That seemingly innocent observation shifted the direction of her Ph.D. work – and of her life.

If you cut a human brain straight down the middle between its left and right sides, the knife will hit only a relatively small amount of tissue that connects the otherwise separate hemispheres. The largest portion of this connecting tissue is the corpus callosum. Sliced through and viewed from the side, the corpus callosum looks like a tall, skinny 'C' lying face down. It is about three inches from end to end, an inch and a half high and only about half an inch thick, although the thickness varies from one spot to another.

The end of the 'C' that sits closest to the back of the head is called the splenium, and it was in this area that de Lacoste spotted a very noticeable sex difference. In each of the five female brains, the splenium was more bulbous than in the male brains.

Fascinated, de Lacoste went back to the brain specimens and made a series of careful measurements. They verified what her eyes had told her. The females' spleniums bulged out more than the males', they were wider and they had a larger area on average. And, when she adjusted for brain weight, the entire corpus callosum was larger in the females than in the males.

By itself, this would be a noteworthy discovery. The corpus callosum has nothing to do with sexual or reproductive function but rather plays a role in intellectual processes, so a sex difference here would seem to imply an innate difference in how men and women think. But there was more to it than that. De Lacoste's finding appeared to support a controversial theory about why men excel in spatial ability and women in verbal skills, and so her report created a sensation.

Scientists have known for some time that the two sides of the brain are specialized for different functions. In most people, the left side is the dominant hemisphere for listening, speaking, reading and writing, while the right hemisphere does much of the work in spatial thinking and problem solving. And in the 1970s psychologist Jerre Levy and others suggested that this division of labor may be somewhat different in males and females. At the heart of Levy's theory was the 'cognitive crowding hypothesis': the idea that if one hemisphere tries to handle two different types of cognitive abilities at once, they will interfere with each other. And since verbal functions are more important to humans than spatial reasoning, she suggested, it will be the spatial skills that are hindered more by this crowding. Working from there, Levy hypothesized that the right hemisphere in men is more highly specialized for spatial reasoning than it is in women, making men's spatial skills less vulnerable to cognitive crowding from verbal processing and thus providing men an advantage in spatial abilities and mathematics. On the other hand, the right hemisphere in women would play a larger role in language processing than it does in men, giving women more brain power devoted to verbal skills and imparting them an advantage in this area.

The theory was controversial because it implied a biological – instead of an environmental – reason that men and women have different patterns of mental abilities. The scientists and others who insisted that there are no such biological differences pooh-poohed Levy's idea, when they weren't studiously ignoring it.

But there should be ways of testing the theory. If Levy was right that men's and women's brains divide up tasks between the hemispheres differently, then the hemispheres should have different communications needs in the two sexes. The right and left sides of a male brain, being more specialized, would not need to talk to each other as much as the two sides of a female brain, which did more similar jobs.

And that's exactly what de Lacoste's finding seemed to imply. One part of the corpus callosum – the splenium – appeared to be larger in women than in men, and a larger corpus callosum should mean more communication between the two halves of the brain. At the very least, the mysterious observation that the splenium was more bulbous in women implied that the corpus callosum was designed differently in men and women, which might be related to a difference in how the two sides of the brain communicated. De Lacoste's paper, published in the prestigious and widely read journal *Science*, got a tremendous amount of attention from newspapers, magazines and television news, and de Lacoste, still a graduate student, suddenly found herself notorious. She remembers meeting one researcher at a meeting who said, 'Oh, so you're the infamous neuroscientist.'

Much of the initial reaction from scientists was favorable, de Lacoste says, but she soon discovered that she had signed up for a full-contact sport. A vocal minority of scientists were outraged that her results had been published and given so much attention, arguing that such sex differences might be misinterpreted as proving that women have inferior brains and could be used to justify discrimination against women. Of course, neither de Lacoste nor any other scientist ever suggested that female brains were inferior – just different – but that didn't matter. 'I went for a job interview at one university,' de Lacoste recalls, 'and while I was giving a talk on my work, a man stood up in the audience and said, "How dare you do research on sex differences in the brain?"'

After de Lacoste's findings were made public, other scientists began to look at the corpus callosum, seeking to study the sex differences she had found, and it soon became clear that they weren't as obvious as they had seemed from her paper. Some researchers saw no differences at all, others saw differences but not the same as de Lacoste's, and by 1985 the tide of scientific opinion started to turn against her. Although scientists are still debating the existence of sex differences in the corpus callosum (more on this a little later) and de Lacoste could still turn out to be right in the end, she found herself exhausted by defending her results and is now working in another area. 'I decided after all that,' she says, 'that sex differences is too volatile a field.'

Looking back on it, de Lacoste says she wishes she had done a few things differently. It would have been better, she admits, to have waited and gotten data on more than five female and nine male brains, but at the time the evidence seemed so clear-cut that fourteen brains appeared to be enough. And it wasn't a good idea to publish her paper in the high-profile *Science*, she says now, because the resulting publicity politicized her results and for ever associated her in people's minds with that one finding. 'I believe it didn't help my career at all,' she says, and it might have hurt it. Still, she continues to feel the attraction of studying sex differences and perhaps, she says, she'll get back into the field in five or ten years after things have settled down.

Given the contentiousness of the subject, it's surprising to meet the researcher who has done more than any other to pin down the existence of physical, measurable differences between men's and women's brains. It is not an iconoclastic old curmudgeon, looking to prove women inferior. It is not a full professor, comfortable in his tenure and looking to stir things up. It is not even an outspoken, politically incorrect Young Turk looking to get his face on magazine covers. She is a recent Ph.D., aware that most people shy away from such questions and sensitive to the political overtones, but quietly determined to get some answers.

Laura Allen is petite and reserved, with light hair and blue eyes that look straight into yours as she speaks, always very politely and very softly. During our first interview, in a lab at the University of California, Los Angeles, she talked so quietly that my voice-activated tape-recorder failed to record much of the interview because it thought no one was speaking. At the time she was six months pregnant with her second child, a girl to accompany a boy born three years earlier, and she calmly downplayed any negative reactions her work had provoked. 'I think maybe people involved in the women's movement are very reluctant, very hesitant about some of this work, [fearing] some of the data may be used against them,' she says, but she doesn't believe there's any basis for these fears. 'I think that if the public is educated, the results won't be misinterpreted.'

To date, Allen has found sex differences in five separate places in the human brain – which might not seem like an impressive number considering how big the brain is, but just give her time. After all, she has performed her precise, time-consuming measurements on only seven spots so far, so she's had a pretty good success rate. 'And we expect to find many more differences as we look in more places,' she says.

If Allen is right, the brain may be riddled with quite obvious sex differences, if only one knows where and how to look. It is an idea that would have been greeted skeptically a decade ago, when de Lacoste's findings on the corpus callosum were new, and that would probably have been dismissed completely a decade before that, when the modern era of research into brain sex differences began.

It was in 1971 that two British researchers, Geoffrey Raisman and Pauline Field, reported the first structural sex difference in the brain of a mammal, and it had not been easy to find. The difference was not in a human but in a rat, and it was so subtle that it could only be seen with the aid of an electron microscope – a very powerful microscope that employs electrons instead of light and can see much smaller details than a light microscope. Examining a part of the rat's brain involved in regulating hormone production, including the timing of ovulation in females, the two researchers focused on its synapses – the points where two neurons make contact. The synapses were of two types, they saw, and when they counted up the synapses in males and females they discovered a difference: the ratio of the 'type I' synapses versus 'type II' synapses was consistently larger in male rats than in female rats.

Granted, this was a rather mysterious sex difference and Raisman and Field didn't have any idea what it meant, but it was a start. For some time, scientists in the field had been expecting to find physical differences between male and female brains – after all, male and female rats behaved very differently, especially when it came to sex and reproduction, and something in the brain must be responsible. Researchers could even say with conviction that the brain differences must be caused by sex hormones during development, since exposing rats' brains to the right hormones could make the rats behave like members of the opposite sex.

But at the time researchers had no real idea of what the sex differences in the brain would look like. It seemed likely that concrete differences would be very hard to find, and perhaps they would all be as subtle as the first one. It took only seven years to prove otherwise.

In 1978, Roger Gorski at UCLA (whom we met in Chapter 3 showing rat porno movies) found a sex difference in the rat brain that wasn't subtle at all. Laura Allen, who was an undergraduate working with Gorski on an honors research project at the time, remembers Gorski's entire lab being exhilarated by the finding. Here was a contrast that practically jumped out at you. No need to count synapses or even to depend on the electron microscope – a light microscope at low magnification was plenty. And it was in the same general area of the rat brain that Raisman and Field has seen their difference. Apparently they had been so busy peering at the trees that they had missed the fact that the forest was several times bigger in males than in females.

It was actually a post-doctoral student in Gorski's lab, Larry Christensen, who first stumbled across the difference while studying slices of rat brain under the microscope. In the hypothalamus, an area involved in controlling sexual activity, he noticed that a particular 'nucleus' (a group of interrelated neurons) was visibly larger in males than in females. After he reported it to Gorski, practically the whole lab dropped what they were doing to follow up on it. 'It was one of the few times, if not the only time, that I was paranoid about getting something published,' Gorski remembers. The difference was so obvious that he was sure somebody else would notice it too and publish first. No one else did, though, and Gorski was able to lay claim to it. After measuring the regions in a number of rats (are were two to a brain, one on the right and one on the left), Gorski found that they were consistently three to seven times larger in males. He named the area the sexually dimorphic nucleus, or SDN. (Christensen, after his serendipitous discovery, eventually chose to leave science and become a Buddhist monk, Gorski reports.)

Two years later, Gorski proved that the size of the SDN is determined by what hormones the rat is exposed to around the time of birth. If he injected female rats with testosterone, their brains grew an SDN the size of normal males, and if he treated male rats with hormones that block the action of testosterone, the SDN got no larger than a normal female's. The same thing that made a rat behave sexually like a male inflated its SDN to a male-like volume.

It was a breakthrough finding – the first evidence in a mammalian brain of a major anatomical difference between males and females. (Two years earlier, a sex difference had been discovered in the size of the part of a songbird's brain that controls its song, but mammals and birds are only distantly related and a similar finding for rats or man had seemed unlikely.) And Allen was in the right place at the right time, getting a chance to work in this exciting new area while still an undergraduate. She had been looking for a research project and knocked on Gorski's door shortly after the SDN's discovery, although she hadn't known of it at the time. Gorski put her to work measuring estrogen uptake in the SDN – research that was never published but which was Allen's first step toward a career in sex difference research.

It was several years, however, before she took her second step. After getting a bachelor's degree in biology in the spring of 1978, Allen spent five years working at UCLA as a research associate performing studies on aging before she returned to school in 1983 to pursue a Ph.D. Initially she had planned to stay in aging research, but she changed her mind after attending a lecture by Gorski in which he described the brain differences that had been found between male and female rats. 'I kept wondering, What happens in humans?' she says. Allen approached Gorski and told him she wanted to study human brains – to look for an SDN in humans.

Gorski agreed to take her on as a graduate student, but he didn't like the idea of her studying humans. Part of it, he says, was his belief that graduate students should cut their teeth on rats before they tackled the much more complicated human brain. But a bigger reason for advising her to stick with rat brains was his worry about how research into the human brain would be received, especially if she did find a sex difference. 'I was worried about sensationalism and how it would affect a young person starting out in science,' he says. The paper on the corpus callosum had appeared the year before, and although Gorski says he wasn't specifically thinking about de Lacoste's experience, it was clear that research into human sex differences wasn't nice, sedate science as usual. Allen should practice safe sex research, he recommended.

Allen, however, was undeterred. Her idea was simple: if the SDN is involved in determining sexual behavior in rats, then there should be something similar in humans because men and women, like male and female rats, differ in their sexual behavior and preferences. Soon, with Gorski's grudging approval, Allen was slowly collecting brains from a hospital.

The obvious place to look for such a sex difference was in the part of the human brain corresponding to the area of the rat brain that contains the SDN. So Allen began her search in the preoptic area and the hypothalamus, two small parts of the human brain believed to play a role in human sexual and reproductive behavior. Following Gorski's lead she combed these areas for nuclei – groups of interconnected neurons that show up as tiny spots in samples of brain tissue when the tissue is sliced, stained and examined under a microscope. (Staining simply refers to the technique of applying a chemical to the brain tissue that makes some nerve cells darken and stand out from the rest of the sample.)

After examining brains from a couple of dozen men and women, Allen found that there were four small nuclei in the area she had chosen. She called them INAH-1, 2, 3 and 4 (for 'Interstitial Nuclei of the Anterior Hypothalamus'), and when she compared their average sizes in men and women, she found that INAH-2 and INAH-3 were two to three times as large in males as in females.

Discovery! It was the type of find that scientists work years or decades for and sometimes never get, and Allen had it in her second year of graduate work. No one else had ever published such a sex difference in the brain – even de Lacoste's sex difference in the corpus callosum was not nearly so striking – so it would be a major coup. But Gorski was hesitant to go public with it. 'He thought it would be so controversial, that there would be an uproar,' Allen remembers. Gorski told her he had heard that a group of scientists in the Netherlands had found a similar sex difference. Why not let them publish first and run interference?

So they held off, and in 1985 Dick Swaab and E. Fliers of the Netherlands Institute for Brain Research reported in *Science* that they had found a nucleus in the human preoptic area that was two and one-half times larger in men than in women. This was the human version of the rat SDN, they suggested, and must play a role in male sexual behavior. Allen and Gorski waited for the controversy, for the uproar – and it never came. They had held off for nothing.

Perhaps the reason for the lack of controversy, Allen says in retrospect, is that the nucleus Swaab and Fliers found lies in a part of the brain involved in sexual behavior, and people expect sex differences there. It wasn't like de Lacoste's report, which involved a thinking part of the brain with no known role in sex or reproduction. Whatever the reason, the muted reaction unleashed Allen.

First she published her findings on INAH-2 and 3, the two nuclei that were larger in males than females. They were in the same area as the nucleus that Swaab and Fliers had found, but were different from it.

Then, thinking back to de Lacoste's findings on the corpus callosum, Allen turned her attention to the anterior commissure – a second, smaller band of tissue that also connects the brain's two hemispheres – and discovered it is an average of about twelve percent larger in females than in males. She looked at the massa intermedia, an even smaller connecting structure between the two sides of the brain, and found it to be seventy-six percent larger in females. And she examined a structure called the bed nucleus of the stria terminalis, which in rats is involved in regulating sexual behavior and aggression, and once again found a sex difference – one part of it was two and a half times larger in men than in women. Out of four nuclei and three other pieces of the brain, five of the seven showed some difference between the sexes.

Allen also took a turn at the dispute that had grown up around de Lacoste's findings on the corpus callosum. Although a number of other scientists had failed to replicate the results, none had done the study in exactly the same way as de Lacoste, who had performed quite precise measurements of different aspects of the corpus callosum's size and shape.

To check it out, Allen decided to look directly at living brains with magnetic resonance imaging, or MRI, which uses magnetic fields to take detailed pictures of the brain. MRI's measurements are not quite as accurate as those done directly on the physical brain at autopsy – although they're getting close – but they have the advantage of being from the brains of live people whose age and other vital statistics can be easily obtained. Borrowing a collection of MRIs of healthy patients from a local hospital, Allen measured the corpora callosa of twenty-four children and one hundred and twenty-two adults. After analyzing the data, she verified one piece of de Lacoste's original finding: the back fifth, the splenium, was wider and more bulbous in women than in men. Allen also discovered that the area of the corpus callosum declined slowly but steadily after age twenty, which might explain the inconsistency of studies on the corpus callosum – if researchers aren't careful that the males and females in their sample are the same age, they may get a misleading result.

From this work, Allen says she believes that the corpus callosum is different between the sexes, although not so strikingly different as de Lacoste first reported. In de Lacoste's original collection of brains it was possible to identify the females simply by looking at the pictures of the callosa and picking out the five with the most bulbous ends. In her own study Allen didn't see such an absolute distinction – some men had spleniums more bulbous than some women – but on average there was still a sex difference. Furthermore, Allen didn't see any sex difference in the size of the corpus callosum or the splenium – the part of de Lacoste's findings that had been the most provocative.

Others get varying results, and the case of the corpus callosum remains confused. Some researchers have found no difference in either size or shape. Sandra Witelson of McMaster University in Hamilton, Ontario, a highly regarded sex difference researcher, thinks there is a difference between males and females but not where de Lacoste saw it. Instead, the part of the corpus callosum directly in front of the splenium – the isthmus – is larger in women than in men, she says. For now, the most that can be said about the corpus callosum is that it may be shaped differently in males than in females, with the splenium in women being more bulbous, but don't bet your life savings on it.

After nearly a decade of looking for sex differences in the human brain, Allen remains nearly alone in the field. Swaab, the Dutch researcher who beat Allen to the punch in 1985, announced in 1991 that the suprachiasmatic nucleus, a part of the hypothalamus which helps regulate circadian rhythms and reproductive cycles, differs in shape between the sexes. It is longer and thinner in females, more rounded in males. And at the Salk Institute for Biological Studies in San Diego, Simon LeVay examined INAH-2 and 3, the two nuclei that Allen found to be larger in men than women. Although his real interest lies in differences between heterosexual and homosexual men (which we'll describe in the next chapter), LeVay replicated Allen's finding that INAH-3 is about twice as large in (heterosexual) men as in women. He did not, however, find a size difference in INAH-2, probably because he didn't study women before puberty or after menopause, the only two age groups for which Allen saw a size difference.

All this is just the beginning, Allen believes. She and the others have looked at only a very small portion of the brain, and although those sections have been the ones most likely to have sex differences – usually parts of the brain involved in sexual behavior – she's doesn't believe the sex differences will stop here. 'There may be a whole region of the cerebral cortex that is sexually dimorphic,' she says, referring to the portion of the brain that is responsible for higher thinking functions. If there is, then eventually someone's going to find it, and it might as well be her.

So the answer to the question posed at the beginning of the chapter – 'Are men's and women's brains different?' – is 'Yes'. On average, some parts of the brain are larger in women, others in men. Some pieces are shaped differently in males and females. More and more differences will probably surface over time, according to Laura Allen and other researchers who study such things.

In some ways the findings are revolutionary. They prove that there are biological sex differences in the human brain, dissimilarities that are probably caused by the different hormone levels males and females are exposed to during development. Although there is no direct proof that hormones are the cause in the human case, we do know that hormones are responsible for similar differences in rats, so it's not a bad assumption that the same thing works for people.

Furthermore, many of the sex differences in brain structure are in just the right places to help explain why men and women differ in such things as sexual behavior and aggression. Laura Allen's INAH-2 and 3 and Dick Swaab's human SDN both lie in a part of the hypothalamus that influences maternal behavior, sexual behavior and the secretion of gonadotropins, hormones that stimulate the gonads (testes or ovaries). The bed nucleus of the stria terminalis, in which Allen has found another region larger in males than females, helps regulate aggression, sexual behavior and gonadotropin secretion. And Swaab's discovery that part of the suprachiasmatic nucleus is shaped differently in men and women may help us understand such sexual rhythms as a woman's menstrual cycle and a man's seasonal cycle of rising and falling testosterone levels. The suprachiasmatic nucleus is thought to be the clock in the human brain that keeps track of twenty-four-hour cycles as well as the passing of months and seasons.

And Allen points out that the sex differences in the sexual parts of the brain are generally much larger than those in the parts of the brain that deal with non-sexual functions. This is exactly what you would expect, she says, since the differences in sexual behavior are much larger than the differences in mental and psychological traits.

On the other hand, the discovery of these few physical sex differences in the brain is no more than a small first step toward tying variation in how men and women behave to dissimilarities in their brains. Nobody has shown that a larger or differently shaped bundle of neurons translates into a measurable behavioral difference, and until someone does, we can only guess as to the true importance of these structural sex differences. Scientists are a long way from being able to point to a piece of the brain that is smaller in men than in women and saying, 'This finally explains why men can't learn to pick their socks up off the floor.'

Indeed, even in rats it has proved exceptionally difficult to figure out exactly what a given piece of the brain is there for. Take Gorski's sexually dimorphic nucleus. It's an average of five times larger in males than females, which makes it the largest known sex difference in a mammalian brain, but Gorski says, 'We've been studying the SDN for fifteen years, and we still don't know what it does.' Because the SDN lies in the part of the rat brain responsible for emotions and instinctive behavior, the obvious guess is that it controls male sexual behavior, but male rats whose SDNs have been destroyed copulate like normal rats, so the answer is not as simple as that. It is perhaps the most frustrating part of his career, Gorski says, that the nucleus he discovered and named a decade and a half ago remains a cipher to this day.

There seems to be only one sex difference in mammalian brains whose function has been exactly pinpointed. It is not in humans, not even in rats, but in gerbils. Over the past several years, Pauline Yahr at the University of California at Irvine has mapped out a small region of the gerbil brain she calls the sexually dimorphic area, or SDA. It lies in the same part of the brain as Gorski's SDN, but its sex difference is a matter of shape, not size. The SDA in male and female gerbils is about the same size, but examined under a microscope it looks quite distinct in the two sexes.

Yahr has proved that in males the SDA controls the sex act. If the area is damaged, a male gerbil will not copulate with even the most willing of females, but destroying nearby regions of his brain leaves him perfectly potent. Yahr hasn't yet proved what the SDA does in female gerbils, but it's probably related to sexual activity as well.

Unfortunately for scientists but fortunately for everybody else, it's not possible to do the same thing with people (university ethics committees tend to frown on the idea of destroying a small bit of a human's brain and seeing what happens), and this makes it even harder in people than in rats or gerbils to learn the precise function of a particular section of the brain. This has slowed down the researchers who are trying to understand how the brain differs between men and women, but it has not stopped them.

One way to get around this difficulty is to learn as much as possible about the brain structure in lab animals, in hopes of being able to draw some inferences about the human brain. Yahr, for instance, has traced out seventy-five different pieces of the gerbil brain that are connected to the SDA. By understanding how this one sexually dimorphic part of the brain fits in with everything else, she hopes not only to discover other pieces of the brain that differ between the sexes but also to assemble a detailed picture of how the gerbil brain controls sexual behavior.

At the University of Illinois, Janice Juraska is pursuing a similar path in studying the corpus callosum of the rat. Like the human case, the rat corpus callosum is different in males and females but, also like the human case, it's hard to pin down exactly what the difference is. Some researchers, such as Victor Denenberg at the University of Connecticut, have reported that the rat's corpus callosum is larger in males than females. Juraska, on the other hand, finds that on average her male and female rats have the same size corpus callosum. It's possible, she says, that the disagreement between her results and Denenberg's arises because they use different strains of rats, or perhaps even because of an environmental difference in how the rats are raised.

But the size difference doesn't really interest her, anyway. 'Size is only a conglomeration' of all the things going on at a smaller scale, she says, and the truth is in the details.

So Juraska has looked at those details – and found a sex difference unlike anything that anyone had expected. Much of the corpus callosum consists of axons (the long appendages on nerve cells which transmit signals to other nerve cells), and the function of these axons is to carry messages from neurons on one side of the brain to neurons on the other side. When Juraska used an electron microscope to examine the axons of the corpus callosum in males and females, she saw that they differed in both number and size. Axons come in two kinds – myelinated, or covered with a thin coating that helps nerve signals travel along the axon, and unmyelinated. Female rats, Juraska found, have more unmyelinated axons than males and sometimes more myelinated axons as well, but males counter by having larger axons.

Since signals travel faster along larger axons, it seems that the male rat's corpus callosum contains fewer, faster axons while the female's has more but slower axons. What does this mean? There's only way to find out: trace out in tedious detail just where these axons go. Cutting through the corpus callosum is like slicing through a bundle of wires in a large telecommunications system – you can see the individual wires but you don't know what any of them are connecting. So Juraska and her students are creating a map that will describe the function of each bit of the rat corpus callosum. These axons connect the visual cortex on the right side with the same area on the left side, those run between the motor areas of the two hemispheres, and so on. When it's done they will be able to look for patterns in how the axons vary between male and female rats. Perhaps, for instance, the fewer, faster axons that males have give them faster muscle reaction times, while the more numerous, slower axons of females offer greater fine control of muscle movement.

Unfortunately, it's next to impossible to analyze human brains with the same detail that Yahr and Juraska are providing for rodent brains. In order to get such details, Juraska says, she must prepare her rats' brains very carefully and start the treatment soon after the rat is killed, before its brain has time to deteriorate. 'The common wisdom is that anything over twenty seconds is bad,' she says.

Doing the same thing with a human brain would mean waiting around a hospital for someone to die and immediately removing the brain and putting it in a solution to keep it from degenerating – not something that any researcher wants to do, or that hospitals or next of kin would tolerate. So the post-autopsy human brains that researchers get from hospitals arrive in various states of deterioration. They're generally OK for measuring sizes or looking for large features (although even then researchers must worry about how much the brain may have shrunk), but many of the fine details have become muddied.

There is, however, a relatively new approach that gives scientists a whole new way of looking for sex differences in brain structure. With magnetic resonance imaging, the technique that Laura Allen used to measure the size of the corpus callosum in men and women, researchers can look into the brain of a living person. Although MRI doesn't reveal fine details like individual axons (yet), it does give researchers the chance to correlate brain structure with behavior and mental traits. Perhaps the best example of this technique is a study done by Melissa Hines and coworkers at UCLA in 1991. They gave a series of tests to twenty-eight women volunteers and also put each of them in an MRI unit to get pictures of their corpora callosa. In this way Hines could perform a direct test of the concept hinted at by Christine de Lacoste's controversial paper of a decade earlier – that the structure of the corpus callosum is related to verbal skills or other mental abilities, and in particular that a larger splenium (back fifth of the corpus callosum) is responsible for women's superior verbal abilities.

When Hines analyzed her data she discovered a pattern nearly as remarkable as de Lacoste's male/female difference in the splenium. (Surprisingly, the study got almost no press attention, probably because the paper is rather technical and it appeared in a journal that few reporters read.) On average, the women subjects with larger spleniums performed better on such verbal tests as coming up with words that begin with a certain letter and thinking of synonyms for a given word. Furthermore, Hines found a relationship between the size of the splenium and 'language lateralization' – how much the use of language is isolated in one half of the brain as opposed to being spread across both hemispheres. Women with larger spleniums tended to have their language processing divided more evenly between the two sides of the brain. In short, Hines's experiment supports the idea that a larger splenium is correlated with better language skills, and that these better language skills arise when the language-processing areas in the right and left hemispheres talk with each other more.

Because she wanted to keep her experiment simple Hines did not include any men, so it's impossible to say how her findings fit in with the sex differences in the corpus callosum that some researchers have seen (and others haven't). But the pieces of the puzzle are all there: women have better verbal skills than men on average; the splenium seems to be different in women and men, in shape if not in size; and the size of the splenium is related to verbal ability, at least in women. It will just take more tests and more observations and more arguing to decide how all the pieces fit together.

And that's the most frustrating part of the study of structural sex differences in the brain. Researchers like Laura Allen have proven that the differences exist and that they lie in parts of the brain which influence mental and psychological behaviors different in men and women, but so far (except for one piece of the gerbil brain) nobody has made a direct connection between structure and behavior. It's frustrating, at least, for outside observers like you and me who just want to know the bottom line. But for the scientists it's a challenge – they've got questions that'll keep them busy for decades.

Size and shape – all of the structural sex differences found so far in the human brain involve one or the other, or both, of these simple attributes. They're easy to spot with either post-mortem physical examinations or MRIs, but they're surprisingly hard to interpret. Shape is an enigma – some researchers speculate that different shapes imply the neurons are connected differently, but that's no more than an educated guess right now. Size seems like it would be easier to decipher – bigger means more neurons, or at least larger neurons, which means more brawn in that part of the brain, right? – but even that is deceptive.

The largest and most mysterious of the sex differences in the human brain is a size difference. It has been known about for more than a hundred years, but even today there is no agreement on what it means. It is seldom discussed in the scientific journals and almost never mentioned in the popular media, partly because no one knows quite what to make of it and partly because of a worry that this major difference will be misinterpreted and misused, as it was a century ago. Its story offers a valuable lesson for interpreting sex differences in the brain.

In the middle and late nineteenth century, craniometry – the measurement of skulls and brains – was all the rage. Doctors and scientists competed with one another to get the most interesting brains for their collections, and the most prized of all were those that had once occupied the heads of brilliant scientists and intellectuals. In reading about the time, one gets the impression that these men of science could hardly wait for the body to cool in order to pluck out the brain, weigh it, measure it and add it to the pile.

Much of this fascination arose from the conviction that a person's intelligence was closely related to the size and structure of his brain. It may have seemed like a logical belief, but as Stephen Jay Gould points out in *The Mismeasure of Man*, scant evidence for that theory could be found in the measured sizes of great men's brains. (The idea of a 'great woman' was, of course, still decades away.)

One of the biggest brains the craniometricians found belonged to the Russian novelist Ivan Sergeevich Turgenev, whose brain weighed over two thousand grams, about fifty percent more than average, or exactly what was expected from such a brilliant gent. Yet, on the other hand, the brain of the Nobel Prize-winning French novelist Anatole France weighed barely half of Turgenev's and was much smaller than average. The French zoologist Georges Cuvier, a founder of comparative anatomy and paleontology, impressed everyone with his 1,830-gram mark, but it was rather embarrassing to discover that Franz Josef Gall, the German anatomist who claimed that mental capacity could be read in the shape of a person's skull, had a brain that weighed only a below-average 1,198 grams. And how could one explain the fact that criminals, the dregs of society, sometimes proved to have larger brains than its most prominent and celebrated members? Gould writes that the assassin Le Pelley had a 1,809-gram brain that rivalled Cuvier's, and one woman who murdered her husband scored a noteworthy 1,565 grams.

Still, the craniometricians' conviction that brain size determines intelligence was strong enough to overcome the somewhat contradictory evidence of the collected brains, and with a little cleverness these men of science found it possible to explain away the apparent inconsistencies. Why was the brain of this genius so small? Well, he was quite old when he died and his brain must have atrophied a lot since he was in his prime. And what about this intellectual's rather small skull? He was short, and since everyone knows that brain size is related to body size, he certainly would have had an even tinier head if he hadn't been so smart. As for the criminals, perhaps it took nearly as much intelligence to lead a successful life of crime as to make your mark on the right side of the law. There's always an explanation if one looks hard enough.

The assumed connection between brain size and intelligence offered the craniometricians an 'objective' way of comparing the intelligence of different groups of people. Unfortunately, these comparisons were no more objective than their analyses of the brain sizes of intellectuals.

Such comparisons were a favorite pastime of Paul Broca, a famous French surgeon and the discoverer of Broca's area, a part of the brain that controls speech. As Gould describes him, Broca was a careful, serious scientist who spent a great deal of time perfecting his measuring techniques. He worked out, for instance, a way of packing lead shot into a skull to gauge the size of the brain that inhabited it and proceeded to use this method for studying brain sizes of people in centuries past. Broca's one weakness was that too often he was sure he knew the answer before he had even started to look at the evidence.

This was the case when Broca compared men's and women's brain sizes, according to Gould's telling of the tale. Based on data from hundreds of autopsies done at Paris hospitals, Broca calculated that the average male brain weighed 1,325 grams, while female brains averaged only 1,144 grams. The data appear to have been impeccable, Gould says, but the way that Broca interpreted them was not. From earlier studies Broca had calculated that there was a correlation between body size and brain size – he had even adjusted the numbers in other work to take this into account – but when he published his data comparing brain sizes between the sexes, he never bothered to adjust for the fact that women have smaller average body size. There was no need to go through all the calculations, he said, since it was a foregone conclusion: women are less intelligent than men, so obviously their brains must be smaller. Case closed.

Such 'scientific' analyses were sometimes used to justify treati. women as inferior creatures. Indeed, Gustave Le Bon, one of Broca's disciples, put the argument very clearly in a statement remarkable for how well it illustrates the ugly, condescending attitude of some of the men of that time:

> [T]here are a large number of women whose brains are closer in size to those of gorillas than to the most developed male brains. This inferiority is so obvious that no one can contest it for a moment; only its degree is worth discussion. All psychologists who have studied the intelligence of women, as well as poets and novelists, recognize today that they represent the most inferior forms of human evolution and that they are closer to children and savages than to an adult, civilized man. They excel in fickleness, inconstancy, absence of thought and logic, and incapacity to reason. Without doubt there exist some distinguished women, very superior to the average man, but they are as exceptional as the birth of any monstrosity, as, for example, of a gorilla with two heads; consequently, we may neglect them entirely.

With such sentiments, the physical evidence hardly mattered. Whatever the data said, they were sure to be interpreted as proving that women were inferior.

In an ironic footnote to this tale of brain size and sex, Gould describes how Maria Montessori showed that two could play at Broca's game. Montessori, the teacher and educational researcher whose 'Montessori method' is used in many schools today, argued in the early part of the twentieth century that it is actually women who have the superior brains. Yes, their brains are smaller on average, she admitted, but men have much more muscle mass, whose control demands more of the brain, and if this difference in muscle proportion is properly taken into account women actually have more brain left for intellectual endeavors. Thus females are the smarter sex, she suggested, and it has only been the male's greater strength that has allowed men to dominate women. Take that, Le Bon.

Today, the numbers haven't changed much. The generally accepted figure is that men have brains approximately one hundred and thirty to one hundred and forty grams (about five ounces) heavier than women's brains. What has changed is the interpretation. Nobody claims that women's brains 'represent the most inferior forms of human evolution'. Indeed, as we saw in Chapters 1 and 2, although men and women differ in their patterns of mental abilities, it's really not possible to label one sex or the other as 'smarter'. So why are men's brains so much bigger?

For the past couple of decades the most widely accepted explanation among scientists has been that the difference in size is merely a reflection of men's larger bodies. Broca calculated a weak correlation between body size and brain size, with taller, larger people having – on average – bigger heads and larger brains. The reasons for it aren't clear – presumably it has something to do with more brain cells being needed to control a bigger body – but a number of modern researchers have found a similar association. And since men are larger on average than women, they should have bigger brains on average. Some researchers who have performed mathematical analyses of the body/brain size correlation have concluded that yes, it is possible to account for all of the sex differences in brain size as a function of the sex difference in body size.

Recently, however, more and more scientists have begun to disagree. The difference in average body size, they say, can explain only part of the disparity in brain size. For instance, one recent report based on autopsy records of 1,261 people found that when the males' larger bodies were taken into account the average male brain was still about one hundred grams heavier than the average female brain. Another study that used magnetic resonance imaging to measure the brain size of forty male and female college students found the men to have larger brains even after height and weight were factored in.

Furthermore, Sandra Witelson points out that the difference in brain size appears long before there is any difference in body size. Boys' brains start growing faster than girls' brains between two and three years of age, and by age six, when the brain is almost fully grown, the male advantage is already established. 'Thus, brain size is clearly not determined solely by body size,' Witelson says.

It's something that women – and men – must grapple with, says Janice Juraska. It was comforting to assume that men's larger bodies were the reason for their larger brains, she says. 'It makes women feel better: of course their brain is smaller – their bodies are smaller.' But now it's important to find the real reason for the difference.

Juraska has done a lot of thinking about the size disparity since finding a similar difference in rats. In 1991, when she examined the binocular area of the rat visual cortex (the part of the rat's brain that processes visual information and combines data from both eyes to provide depth perception), she discovered that the area is almost twenty percent larger in male rats than females. More importantly, the difference was not due simply to males having larger neurons – the male rats actually have about twenty percent more neurons than the females.

Juraska was shocked. 'I was going around saying, "What do they do with all those neurons?"' Most of the other size differences in the rat, such as the SDN, had appeared in parts of the brain that deal with reproduction and sexual behavior, but the visual cortex is about as close as the rat gets to an intellectual center. Although some parts of it are devoted to taking in raw data from the optic nerve, that's not where Juraska saw the sex difference. Instead, it was in the layers of the cortex where the brain processes that data to develop an image of its surroundings.

The male rat's brain must be doing something different from the female's brain with its visual system, Juraska decided, and initially it was hard for her to shake the feeling that 'different' meant 'better'. Her finding seemed to imply that – in the visual cortex, at least – the male rat had more brain power than the female. 'I was actually depressed by this finding,' Juraska remembers.

Not for long, however. She soon realized that she was making the same mistake the craniometricians did by assuming that bigger inevitably means better. There are actually quite a lot of reasons that could explain why the male rat has a larger visual cortex than the female, she says, and most of them don't imply that the male's brain is any better – just different. Perhaps, for instance, the brain of the male rat uses more neurons to do essentially the same job as the female brain – it's less efficient in some sense. Christina Williams at Barnard College of Columbia University has shown that male and female rats use different strategies to learn their way around a maze, with the male relying on geometric cues and the females preferring landmarks, or perhaps using both equally. If using geometric cues takes up more brain space than the landmark strategy, the male rats might need more neurons to do the same job.

Juraska believes that something similar is going on in the human brain. One possibility suggested by several scientists is that the extra ounces in an average man's brain may be devoted to spatial ability – the one major talent that men seem to excel in. (This hypothesis is independent of, but not necessarily contradictory to, the idea that high spatial ability arises from one half of the brain specializing in spatial skills. Both factors could be at work.) Perhaps, Juraska says, visualizing objects in three dimensions requires proportionately more brain cells than verbal ability. If so, the brain size difference may be nothing more than a reflection of what we already know – that males generally excel in mental rotations, orienting themselves with a map and other spatial skills.

The thing to avoid in thinking about the difference in brain size, Juraska says, is the simplistic idea that brains can be weighed and compared like sacks of sugar. If one sack of sugar weighs five ounces more than a second sack, it's clear that the first one is 'better' in the sense that it can do more – bake more cakes, sweeten more cups of coffee, and so forth. But brains aren't nearly so simple. 'What do you do with a bigger brain?' Juraska muses. 'Do you put in more cells? Bigger cells? Do you put in extra things, or do you get sloppy and less efficient about what's already there?' (David Ankney of the University of Western Ontario reports one suggestion he heard on a radio talk show. Upon hearing a report on Canadian radio that men's brains are about a hundred grams larger, one woman called in and said, 'You know how men have one thing on their minds from the time they wake up to the time they go to sleep. The extra brain cells are there so they can think about sex all the time and still be able to function somewhat normally.')

Recently, a group of doctors and researchers in Iowa performed an analysis of male and female brains which agrees with Juraska's idea that brain size may mean somewhat different things in men and women. The researchers used MRI to measure both total brain size and the volumes of different sections of the brain (the temporal lobes, cerebellum, hippocampus and so on) in sixty-seven men and women, and then they correlated the brain measures with the subjects' scores on an IQ test. Their goal was to see if there is a relationship between brain structure and different types of intelligence, and to test whether that relationship is the same in men and women.

They found small correlations between IQ and brain size. On average, the subjects with larger brains had higher IQs, although there was a great deal of variability with many smaller-brained subjects scoring well and many larger brained subjects scoring poorly. But they also found that the correlations differed between the sexes. In general, a woman's brain size tended to be more closely related to her verbal IQ while a man's brain size correlated better with his performance IQ (a measure that includes spatial ability and mathematical reasoning). The implication is that men's and women's brains are structured differently, so that – in some very rough, average way – a larger female brain is likely to have greater verbal intelligence while a larger male brain is likely to have higher spatial and mathematical abilities. It's important to note that these correlations are relatively weak – brain size accounts for only a small part of the variation in intelligence from one person to another, and the 'quality' of the brain seems to be much more important than the 'quantity' – but still the differences between men and women do imply that their brains handle things differently. Thus, as Juraska said, it makes no sense to make a direct comparison between the sizes of male and female brains.

These results remain speculative and controversial. Indeed, many scientists deny that there is any connection at all between brain size and intelligence, although a number of recent studies have found that a weak correlation does exist. For now, it seems, only one thing is sure. The size difference, as Juraska says, has the same implication as the rest of the sex differences found in the brain: it shows that men's and women's brains are different, but it does not imply that one is better than the other.

CHAPTER 6

Not Quite the Opposite Sex

For years Simon LeVay had labored in relative obscurity, trying to understand how the brain takes signals from the eyes and produces an integrated picture of the world. Although he had the respect of his scientific peers, few outside the small world of visual cortex research had ever heard of him. Then in the summer of 1991 he published a single paper in a field that was new to him and became an instant celebrity.

It had all begun a couple of years earlier when LeVay read an article by Laura Allen. Deep in the hypothalamus, a part of the human brain that governs many aspects of sexual behavior, Allen had found one bit of tissue that was twice as large in men as in women. Reasoning from what was known about rats, Allen speculated that this nucleus, which she called INAH-3, might be involved in sexual orientation – what makes a person attracted to one sex or the other.

And that got LeVay thinking: if INAH-3 does have something to do with sexual orientation, then shouldn't it differ somehow between heterosexuals and homosexuals? He decided to find out. After asking advice from colleagues familiar with the hypothalamus (since it was outside his usual area of expertise), LeVay began collecting brains from hospitals and analyzing the area that Allen had described. He autopsied the brains of heterosexual and homosexual men as well as of heterosexual women, and when he was done he had discovered a remarkable pattern: INAH-3 is large in people sexually attracted to women and small in those attracted to men. On average, the nucleus is twice as big in heterosexual men as it is in either homosexual men or heterosexual women.

When the discovery was published, it triggered a flood of news stories, opinion columns and letters to the editor, and LeVay got his fifteen minutes of fame. Everyone wanted to know: Does this mean that sexual orientation is biologically fixed? Are homosexuals born that way? Many people, particularly some homosexual rights activists, treated the study as proving that homosexuality is innate, not a choice. LeVay, who is himself homosexual, was more cautious and acknowledged that one study cannot settle the issue, but he did take advantage of the attention to argue for greater acceptance of homosexuality.

Lost in the brouhaha over whether homosexuality is right or wrong, inborn or opted for, was the fact that LeVay's finding has implications for all people, not just homosexuals. After all, if this bit of the hypothalamus holds some hint about why homosexual men are attracted to other men, then it might also offer clues about why heterosexual men are attracted to women and heterosexual women to men.

Think of it this way: LeVay's study can be compared to Sheri Berenbaum's research on CAH girls, those females whose brains are exposed to excess testosterone in the womb. By studying why CAH girls would rather play with trucks and Lincoln logs than dolls, Berenbaum hopes eventually to understand why normal boys and girls make different choices at playtime. Similarly, LeVay and other researchers who study people with a reversed sexual orientation are opening a window on two of the great mysteries of sex. Just what is it that women see in men? And likewise, what is it that men see in women?

Although people rarely look at it from this angle, sexual orientation is a sex difference, and it is the second largest of all sex differences. (We'll get to the first in a minute.) Most men are sexually attracted to women, while most women are sexually attracted to men. In the United States, the best current estimate is that at least ninety-five percent of males and more than ninety-eight percent of females are attracted to the opposite sex. As a sex difference this dwarfs the disparities in verbal ability and spatial skills and it far surpasses even the inequality in physical size.

What causes people to fixate on the opposite sex? Or, as many of us have wondered, just why is it that we 'can't live with them, can't live without them'? The first part of that old saying is easy enough to understand, but no one yet has figured out the animal magnetism that makes these two opposites attract. It may be that the best way to answer the question is to study people who *can* live without the opposite sex and to understand what makes them different.

By the way, you've probably read or heard somewhere that about ten percent of the population of the United States is homosexual. This is a number that has been quoted for years, but every expert in the field that

I've talked to says it's wrong, and in the past year a number of stories have appeared in the newspapers and on TV rebutting it. One recent study found that in the United States just 2.3 percent of men reported ever having had sex with another man and only 1.1 percent said they were exclusively homosexual. Of course, these numbers could be low since some people may not want to admit they are homosexuals, and the exact figure will change depending on how you define homosexuality, but the researchers who study the issue tend to agree that probably about four to six percent of American males and about two to three percent of American females have had homosexual contact sometime in their adult lives. If you demand a strict definition of homosexuality, such as having had homosexual sex with someone in the past year, the numbers are probably more like two percent and one percent. Recent sexual surveys in Britain and France found very similar percentages there.

The ten percent figure has become something of a political football. It got its start with the badly flawed Kinsey Report of more than forty years ago, and homosexual rights groups have continued to use it in spite of more recent and better numbers. One such group, for instance, even calls itself 'One in Ten', but the accurate figure is more like one in twenty-five or one in fifty.

Before continuing with sexual orientation, let's take a brief detour to look at the largest single sex difference, larger even than sexual orientation. It is so large that most people take it for granted, as if there were no other way for males and females to be. It is sexual identity, the inner feeling that one is either a male or a female.

Do you feel like a male? Or do you feel like a female? If you have a male's body and were brought up as a male, you almost certainly believe yourself to be male, and similarly for females. But there are exceptions. Some people who are biologically male, who were raised as males, who have male genitals and can grow a beard, and who may even be sexually attracted to females none the less feel strongly that they are women. There are also cases of biological women who are convinced that they are really men. These rare 'transsexuals' often say that they feel as if their brains are trapped in bodies of the wrong sex, and sometimes they go so far as to have a sex change operation to make their bodies correspond with their brains. It doesn't happen often – despite the fact that Geraldo and Donahue and Sally Jesse Rafael seem to feature a 'transsexual of the week' – but it does happen often enough that researchers think of sexual identity as a characteristic independent of a person's biological sex. Although no one has an accurate count of how many people are transsexuals, one study in the Netherlands put the number at about one in every 12,000 males and one in every 30,000 females. Or, to put it another way, 11,999 of every 12,000 males feel like males, and 29,999 of every 30,000 females feel like females, which makes sexual identity the most consistent of all differences between the sexes.

It's hard, I admit, to understand how even one person in 12,000 or one in 40,000 could ignore the evidence of their own bodies and be convinced they are members of the opposite sex. For those who are interested, I recommend *Conundrum* by Jan Morris, an autobiographical tale by a British travel writer who grew up as a male, got married and had children, but eventually gave in to his inner, lifelong conviction that he was a woman and had a sex change operation. You still may not understand exactly what he felt, but you'll come away convinced that there are indeed some people – people who are very sane, rational and otherwise well-balanced – whose innermost sexual identities do not match their bodies.

Researchers are still arguing about how this happens. In the 1960s and 1970s, the most popular theory about sexual identity was that it is a learned behavior. If a child looks male (or female) and is raised like a male (or female), then the child will inevitably develop a male (or female) sexual identity. It's undeniable that society sends a strong message to each child that says either 'You're a boy' or 'You're a girl', and that plus the evidence of a male or female body should be enough to create a permanent male or female identity, or so the theory went.

It made sense. It made so much sense, in fact, that in a few cases where a young male lost his penis by some freak accident, doctors convinced the parents to try to turn him into a girl – surgically reshape the rest of his genitals, put him on female hormones, change his name and raise him just like a girl.

The most celebrated of these cases was a seven-month-old boy – one of a pair of identical twins – whose penis was burned off in a circumcision by cautery that went awry. The current on the electrocautery device was set too high, and his entire penis was destroyed. After long consideration and consultation with experts, his parents decided to make him into a girl since he could never have a normal life as a male. His testicles were removed, the doctors fashioned a vagina for him, and he would be given female hormones. This was in the mid 1960s. In 1972, John Money of Johns Hopkins University, one of the most prominent supporters of the socialization theory of sexual identity, reported that the 'girl' was doing well and had accepted her sex, even if she did have 'many tomboyish traits, such as abundant physical energy, a high level of activity, stubbornness, and being often the dominant one in a girls' group.' It seemed like proof that a person's sexual identity was fixed by society, not biology, and it was widely reported as such at the time. But things were not that simple.

In 1979, Julianne Imperato-McGinley reported her startling study (described in Chapter 3) of a group of males in a Caribbean country afflicted with 5α-reductase deficiency, a genetic defect that made them look like girls at birth but develop male genitals at puberty. Eighteen of these males had been raised completely as girls, she said, but seventeen of them were able to adopt a male sexual identity after their male genitals appeared. The eighteenth continued to live as a woman, wore false breasts and said he wanted to have genital surgery so that he could be a 'normal' female.

Imagine living the first dozen years of your life as a female and then sprouting a penis and scrotum at puberty. It would be a tremendous shock, to say the least, and if the socialization theory of sexual identity were right, you would never be able to shake the conviction that you were a female on whom fate had played a rather nasty trick. Yet of the eighteen males Imperato-McGinley studied, only one behaved this way. The rest shifted to thinking of themselves as male – proof, she said, that something else must be influencing sexual identity, something that must have its roots in the male hormones the boys were exposed to in the womb or at puberty. A dose of male hormones, she said, was enough to help undo a decade of socialization.

Not so fast, replied other researchers who questioned Imperato-McGinley's claims. They argued that it was unlikely the seventeen males had been raised unambiguously as girls – their parents must have known something was amiss by their unusual, pseudo-female genitals at birth – so the boys probably never developed a strong female sexual identity and could therefore shift gears without too much trauma when their sex changed before their eyes. Today, Imperato-McGinley continues to insist that the males were raised as girls, and the argument may never be settled to everyone's satisfaction since most of the subjects are now forty or older and their memories of their childhoods are fading.

Recently other cracks have begun to appear in the socialization theory of gender identity. The boy-turned-girl twin that John Money made famous in the early 1970s wasn't such a success story after all, it now seems. Milton Diamond, a sex researcher at the University of Hawaii who has followed the case, reported a decade later that 'this individual has never accepted the female status or role as claimed by the early investigators'. Before he reached puberty and before he had been told of his sexual history, the transformed twin was having a lot of trouble being a girl – he didn't fit in with his peers as a girl, and a psychiatrist who examined him concluded that he might never grow into being comfortable as a woman. Later, when the girl/boy turned eighteen, he had plastic surgery to rebuild a scrotum and penis so that he could have sex as a male, although he is of course sterile. Now, Diamond says, 'he lives as a male and seeks females as sexual partners.' In this one case, anyway, it was clearly not enough that the child look like a girl and be raised as one.

But the biggest problem for this socialization theory, Diamond says, is the existence of transsexuals. Despite looking like members of one sex, having been raised as members of that sex and being censured by society for their feelings, they still insist that they do not belong to the sex they appear to. If sexual identity is purely dependent on how a person is raised, it's hard to explain people like Jan Morris. And, Diamond notes, a mixed-up sexual identity is more than just an extreme version of homosexuality. Although most female-to-male transsexuals are lesbians who are attracted to women and who take the extra step of thinking of themselves as men, male-to-female transsexuals are not so predictable. Some of them are homosexuals who are attracted to other men and who think of themselves as women, but there is a second group of men who think of themselves as women yet who are at the same time sexually attracted to women. When this type of male-to-female transsexual undergoes a sex-change operation, her new life is as a lesbian. Over the years of working with such people, Diamond says he has found that just about anything can and does happen. In one case he was familiar with, a married couple both decided they were in the wrong bodies, both had sex-change operations, and then stayed married with their roles reversed, the husband now a wife and wife a husband. It's hard to explain such things in terms of environment, he comments, especially since these people seldom report being raised any differently than other people and there is rarely any evidence of cross-rearing – raising a boy like a girl or vice versa.

Diamond is convinced that a person's sexual identity is established by sex hormones biasing the nervous system in the womb and that there is little or nothing that can be done to change it after the person is born. Many other researchers still believe that environment plays a role, but nowadays few contend that biology has nothing to do with it. At least part of our sense of being a man or a woman appears to be imprinted on our brains by our hormones.

The debate over the causes of homosexuality is similar to that about transsexuality. In the past, popular psychological theories have ascribed homosexuality to such causes as a distant father or overbearing mother, but recently an increasing amount of evidence has pointed in another direction. Sexual orientation, it now seems, is strongly influenced by the level of hormones during a critical time in the development of the brain. The question of which sex it is that you can't live without could be mostly a matter of testosterone.

Some of the evidence comes directly from the brain. Although LeVay's work was the most highly publicized, other studies have also found differences between heterosexual and homosexual brains. In 1990, a year before LeVay's report, the Dutch team of Dick Swaab and Michel Hofman announced the first discovery of such a structural difference. It lay in the suprachiasmatic nucleus, or SCN, a part of the hypothalamus involved in regulating circadian rhythms (daily biological rhythms, such as the sleep–wake cycle) and other cycles. The SCN, they said, is nearly twice as large in homosexual men as in heterosexual men. The Dutch results didn't generate as much interest as LeVay's findings would, for a couple of reasons: the SCN seems to have nothing to do with sexual behavior or orientation, and it is the same size in women and men. Without a sex difference in the size of the SCN, Swaab and Hofman could not conclude – as LeVay would – that the brains of male homosexuals are feminized, or more like those of women than of heterosexual men. The most they could say is that homosexuals' brains are different, period.

Laura Allen at UCLA, the discoverer of so many sex differences in the brain, has also looked for dissimilarities between heterosexual and homosexual brains. When she checked the anterior commissure, a band of tissue that connects the two halves of the brain and that is larger in women than in men, she found it to be 'super-feminine' in homosexual men – it's even larger in these men than it is in heterosexual women. In ninety autopsied brains, evenly split among heterosexual men, homosexual men and heterosexual women, Allen calculated that the anterior commissure in homosexual men is an average of eighteen percent larger than in heterosexual women and thirty-four percent larger than in heterosexual men.

It's hard to know what this implies about homosexual brains, Allen says, since no one is sure exactly what the anterior commissure does, but it is not thought to be involved in sexual behavior. It connects parts of the brain that are involved with thinking, which implies that homosexual brains differ from heterosexual brains not only in those sections that control sexual behavior but probably in the thinking regions as well.

No one knows yet what causes these differences. It's possible, for instance, that they are the results of behavior rather than its cause. Perhaps being sexually drawn to women somehow makes INAH-3 larger, or finding men attractive causes it to shrink. Most researchers think this is unlikely, however, since in rats the sexually dimorphic nucleus – which lies in approximately the same spot as INAH-3 – does not change in size after the first week or two of life. After that, even castrating a male rat has no effect on the size of his sexually dimorphic nucleus. People aren't rats, but still LeVay and Allen suspect that the size of INAH-3 is set in the womb.

Which leaves the question: Is INAH-3 the part of the brain that decides whether someone is attracted to men or women? It's not that simple, researchers say. Although on average the nucleus is twice as large in heterosexual men as it is in women and homosexual men, there is plenty of overlap and some women and homosexual men have nuclei larger than those in many heterosexual men. This means there must be more to the story of sexual orientation than the size of one small piece of the hypothalamus. Sex hormones during development probably act on that entire region of the brain, LeVay and Allen say, and not only determine the size of INAH-3 but also make other adjustments that combine to influence a person's likely sexual orientation as an adult.

Besides the differences in the brain, several other lines of research indicate that sexual orientation is set, or at least predisposed, in the womb. Part of the evidence comes from CAH females, the girls and women exposed to extra high levels of testosterone in the womb because of a genetic defect. CAH girls are likely to be tomboys and prefer to play with boys' toys, and they also seem to have an increased chance of being homosexual when they grow up. In one study of thirty CAH women conducted a decade ago, five of the women described themselves as homosexual and another six called themselves bisexual – surprisingly high numbers, given that in a group of thirty women chosen randomly from the general population, you'd expect one or two to be homosexual or bisexual, but certainly not eleven. More recent studies have found a somewhat smaller effect – one reported that twenty percent of the CAH women in the study either had homosexual relationships or wanted them, while another found only about five percent of CAH women to be homosexual. This is still larger than the percentage in the general population, so it appears that extra testosterone in the womb does increase the chance a woman will be homosexual.

A second group of women who seem more likely to be homosexual are those whose mothers took DES, the synthetic estrogen once thought to reduce miscarriages during pregnancy. One researcher reported that women exposed to DES in the womb had more homosexual fantasies, were more often sexually attracted to other females and had more homosexual experience than unexposed sisters and cousins. (It might seem contradictory that a synthetic estrogen could make a woman more masculine in her sexual orientation, but remember that much of the masculinization that testosterone causes in the brain is actually carried out by estrogen, after the testosterone has been chemically transformed into the female hormone. If you expose a female fetus to abnormally high levels of female hormones, it will be masculinized almost as if it had been exposed to testosterone instead.)

In both these examples – CAH women and females exposed to DES in the womb – female brains seem to have been masculinized by unusual concentrations of hormones, with the result that the women are more likely to find other women sexually attractive. But what about men? Can male homosexuality be caused by too little testosterone?

So far the best answer seems to be: maybe. It's true in rats, as researchers have transformed male rats into 'homosexuals' by depriving their brains of testosterone at a critical time. (Keep in mind, however, that sexual orientation in rats is not at all the same thing as in humans. A homosexual male rat is merely one that responds to other males like a female by arching his back, not one that has a sexual preference for males.) And there is evidence that testosterone deprivation might have a similar effect in humans. When some women in the 1950s and 1960s took drugs that were supposed to prevent miscarriage and which had the effect of reducing the amount of male hormones the fetus was exposed to, their male children were more likely to be homosexual than children who weren't exposed to these drugs in the womb. The picture is confused, however, because doctors prescribed many different anti-miscarriage drugs during this time, including DES and other synthetic estrogens, and researchers have a hard time figuring out which mothers took which drugs (and when during their pregnancies) and separating the effects of one drug from the others.

Other researchers have found that they can create homosexuality in male rats merely by subjecting the mother to severe stress. (The usual way is to put her in a plastic tube so narrow that she can't move around and then expose her for several hours to very bright lights.) The reason appears to be that the stress hormones produced by the mother somehow reduce the level of testosterone that her fetuses are exposed to in the womb. And one controversial study claims that a similar thing may happen in humans.

In 1980, Günter Dörner, a researcher in what was then East Germany, reported that German men born during and shortly after the Second World War – presumably a very stressful time – were more likely to be homosexual than those born earlier or later. Then in a later study he asked the mothers of a large number of heterosexual and homosexual males about stressful episodes during pregnancy, such as deaths, divorces and other traumatic experiences. He reported that although only ten percent of mothers of heterosexual males reported such stress during pregnancy, a third of the mothers of bisexuals did, as did almost two-thirds of the mothers of homosexuals. Other scientists, however, have failed to reproduce Dörner's results, and he has hurt his credibility with comments about 'curing' homosexuality by making sure that a fetus gets enough testosterone. So most scientists consider the pregnancy-stress hypothesis in human homosexuality as unproven.

None the less, even assuming that Dörner's work is wrong, there is enough evidence to convince many researchers that sexual preference is shaped by the sex hormones in the womb and that homosexuality is caused – at least partially – by an unusual hormone exposure during development. In a 1987 paper, Lee Ellis and M. Ashley Ames spelled it out: '[S]exual orientation in all mammals,' they suggested, 'is primarily determined by the degree to which the nervous system is exposed to testosterone' and other sex hormones, and the critical period for the human fetus is between the second and fifth months of gestation. In short, if a person (or animal) gets a low dose of testosterone during development, she or he will be attracted to (or at least respond to) males, while high levels of testosterone create a sexual attraction to females.

Some indirect support for this hypothesis has appeared in just the past couple of years. Ellis and Ames argued that if sexual orientation is influenced by fetal exposure to testosterone, then homosexuality should have a large genetic component since it is the fetus's genes – which are inherited from the parents – that mostly control what hormones the fetus sees. This in turn implies that homosexuality should tend to run in families. And that is precisely what several recent studies have shown.

The first really convincing study that homosexuality is inherited was published in 1991 by Michael Bailey of Northwestern University in Evanston, Illinois and Richard Pillard of the Boston University School of Medicine. Bailey and Pillard studied one hundred and sixty-one homosexual men who had either twin brothers or adoptive brothers, and they analyzed how often the brothers were also homosexual. (An 'adoptive brother' is one related to you only through adoption – either you or he, or both, were adopted.) The two researchers found that fifty-two percent of the identical twins of male homosexuals were themselves homosexual, compared to twenty-two percent of fraternal twins and only eleven percent of adoptive brothers.

The fact that identical twins of homosexuals are likely to also be homosexual was not surprising – two earlier studies had found that about half of the identical twins of homosexual men are themselves homosexual – but since identical twins are exposed to very similar environments growing up, it was impossible to say whether their similarity in sexual orientation stemmed from genes or environment. Bailey and Pillard's study went further. By including fraternal twins and adoptive brothers, the two researchers could tease apart the genetic influences from the environmental ones. Identical twins have exactly the same genes, fraternal twins share about half of their genes, and adoptive brothers have no common genes, but they all grow up in a similar environment, so if genes play a role in determining a trait, identical twins should be more alike on that trait, followed by fraternal twins (and brothers), and then adoptive siblings. And that is what Bailey and Pillard found: whether or not a man is homosexual depends to a large degree on his genes.

In 1993 Bailey and Pillard followed up that study with a similar one on lesbians and reported similar numbers. Among sisters of one hundred and forty-seven lesbians, forty-eight percent of identical twins were homosexual, as were just sixteen percent of fraternal twins and six percent of adoptive sisters. The conclusion is the same: among females, genetics plays a large role in determining sexual orientation.

Keep in mind, however, that genes are not the whole story. The flip side of the fact that half of the identical twins of homosexuals are themselves homosexual is that half are not. If sexual orientation were completely determined by genes, identical twins would always be either both homosexual or both heterosexual. So what could make one identical twin heterosexual and one homosexual? It's hard to know, Bailey says. It might be subtle differences in the way twins are raised, but that seems unlikely since psychologists have not been able to find any social factors that predispose boys to become homosexuals. Or the difference might arise in the womb, Bailey suggests. Even if two fetuses have the same genes and share a womb, there may be fluctuations in the hormone levels that send one baby down a homosexual path and push the other toward heterosexuality.

Those paths actually diverge in many ways besides sexual orientation. Researchers have found that homosexuals and heterosexuals differ in a number of mental and psychological traits, which in turn offers further indirect evidence that hormones are at the root of sexual orientation. If homosexuality is caused by an unusual hormone environment in the womb, then the echoes of those hormones should be audible not only in sexual orientation but in other characteristics. In particular, if male homosexuals were hit with less testosterone than normal in the womb, they should be more like women on various sex-differentiated mental traits, such as spatial ability. And this is exactly what sex researchers have discovered.

It starts in childhood. Many studies have asked male homosexuals about their behavior as children, and the consistent finding is that they're more likely than other boys to have played 'like girls'. One typical study asked six hundred homosexual men and three hundred heterosexual men about their sex-stereotyped activities as children. It found that seventy percent of the heterosexual men remembered liking baseball, football and other boys' activities, but only eleven percent of the homosexual men did; conversely, nearly half of the homosexual men had liked typical girls' games such as hopscotch or playing house, compared with only eleven percent of the heterosexual men. Furthermore, a third of the homosexual men said they had dressed as girls or pretended to be female, but only one out of ten of the heterosexual men had.

A common recollection of many homosexual men is that they were called 'sissies' when they were boys. In one large study, half of the homosexual men had been thought of as sissies growing up, compared with only one-ninth of the heterosexual men.

UCLA psychiatrist Richard Green dubbed this the 'sissy boy syndrome', on the basis of a fifteen-year study to see if homosexual men really are more likely to have been sissies in childhood or if they were simply more likely to have those memories once they reached adulthood. In 1968, he chose two groups of boys from four to twelve years old. One group of sixty-six boys were extreme 'sissies': most of them had said they wanted to be girls, seventy percent regularly dressed up like girls, sixty percent played with girl-type dolls and eighty-five percent played with girls instead of other boys. The other group of fifty-five boys were typical males, playing sports, hanging around with other boys and seldom doing any girl stuff. Fifteen years later, Green caught up with two-thirds of both groups. He found that three-quarters of the 'sissy boys' had grown up to become homosexual or bisexual, but only one of the other group of boys had.

Conversely, female homosexuals are generally more masculine than other girls growing up. They're likely to have been tomboys, enjoyed playing sports, disliked playing with dolls and playing house, and wished they were boys. One study of fifty-six lesbians and forty-three heterosexual women found that two-thirds of the lesbians had been tomboys, compared to just one out of six of the heterosexual women. And among those who reported being tomboys, the willingness to play with dolls was a key factor in predicting whether they would grow up to be homosexual or heterosexual: of the lesbians who had been tomboys, two-thirds had refused to play with dolls, but of the heterosexual women who had been tomboys, only fourteen percent had avoided dolls.

If all of this sounds familiar, that's because it is. The studies on CAH girls found very similar things: the girls were more likely to be tomboys, to enjoy sports, and to play with stereotypical boys' toys, such as trucks and Lincoln logs, instead of with dolls. It seems that the things a child likes to do are strongly influenced by what types of hormones he or she was exposed to in the womb.

Given these differences in childhood, you might expect heterosexuals and homosexuals to have different psychological characteristics in adulthood. And indeed there are differences, although they aren't nearly as large as you might expect from the stereotypes of the effeminate homosexual male and the masculine homosexual female.

In 1991, for instance, Brian Gladue at North Dakota State University tested aggressiveness in groups of homosexual and heterosexual men and women. He had each respond to a series of statements such as 'When another person picks a fight with me, I fight back', or 'Whenever someone is being unpleasant, I think it is better to be quiet than to make a fuss'. He then rated each person's physical aggressiveness, verbal aggressiveness, impulsiveness, impatience and avoidance of confrontations. He found that homosexuals of both sexes were slightly more likely to avoid confrontations (not surprising, given that they are an often discriminated against minority), but other than that there were few differences. Homosexual men were just as physically and verbally aggressive as heterosexual men, and the only difference between homosexual and heterosexual women went in the opposite direction from the stereotypes – lesbians were less physically aggressive than heterosexual women.

In an earlier study, Gladue had tested heterosexual and homosexual men and women on scales of masculinity (things like competitiveness, dominance, independence and aggression) and femininity (nurturance, compassion, harm avoidance and so on). Homosexual men were less stereotypically masculine than heterosexual men, but no more feminine, and there was little or no difference between homosexual and heterosexual women on either masculinity and femininity.

It's possible that these psychological differences are products of the environment since, for instance, people expect male homosexuals to be less 'masculine' than their heterosexual peers. No stereotypes predict, however, that male homosexuals should be more like women in such mental traits as spatial skill or verbal ability. Yet that is just what several researchers have found.

In 1986, Geoff Sanders and Lynda Ross-Field of the City of London Polytechnic in London, England, reported a major difference on two spatial tests between homosexual and heterosexual men. In one experiment they tested eight heterosexual men, eight heterosexual women and eight homosexual men on a water-level test, which asked the subjects to draw a line on a tilted bottle to indicate where the water level would be if the bottle was about one-third full. There were ten trials with the bottle at different angles. Afterward, the researchers measured how far off (in degrees) each subject was and averaged the scores from each group. The heterosexual women were ten times less accurate than the heterosexual men – that is, their average error was ten times as large – and the homosexual men scored almost exactly the same as the heterosexual women. In fact, only three homosexual men scored better than the worst heterosexual man.

Then Sanders and Ross-Field recruited a second group of subjects – thirteen each of homosexual men, heterosexual men and heterosexual women – and gave each of them two other tests of spatial ability: another version of the water-level test and a mechanical diagrams test that asked the subject to choose which of several mechanisms made of pulleys, levers or cogs will produce a desired movement. On the water-level test, the homosexual men once again performed like the women and much worse than the heterosexual men. On the mechanical diagrams test, the homosexual men scored better than the women but worse than the heterosexual men.

Four years later, Gladue and a group of researchers in North Dakota repeated the water-level test but included a group of homosexual women as well and also added a mental rotations test. On the water level test, the homosexual men again scored much worse than the heterosexual men, and the researchers found that the lesbians also scored more poorly than the heterosexual women. On the mental rotations test, the heterosexual men scored nearly twice as high as members of the other three groups.

The next year Cheryl McCormick and Sandra Witelson at McMaster University in Hamilton, Ontario, gave spatial and verbal tests to three groups of thirty-eight people – heterosexual men, heterosexual women and homosexual men. On the spatial tests, the heterosexual men scored higher than the heterosexual women, and the homosexual men scored in between. McCormick and Witelson saw no significant differences on the verbal tests between any of the groups.

Not all studies have found disparities between homosexual and heterosexual men, however. When Bailey and Gladue collaborated to test spatial ability in sixty heterosexual and sixty homosexual men, for instance, they saw no significant difference on a mental rotations test between the two groups. This last study is the largest and most carefully done to date, and its results indicate that the mental differences between homosexual and heterosexual males are not as large or clear-cut as those between men and women.

Still, when researchers do see differences in spatial skill or other mental abilities, they usually point in the same direction – male homosexuals are closer to women in performance than to heterosexual men, which is precisely what you would expect if male homosexuals were exposed to lower than normal amounts of testosterone in the womb. This, combined with the evidence from other areas, convinces most researchers that male sexual orientation is somehow influenced by the amount of male hormones the brain is exposed to during development.

The picture for homosexual women is much less clear. Lesbians show no signs of being, on average, more like men. If anything, they seem to be slightly more 'feminine' than heterosexual women in certain ways: they are less physically aggressive and score slightly lower on some spatial tests. Thus, despite the evidence from CAH women, extra male hormones do not seem to explain all or even most female homosexuality, and researchers believe that there are probably other factors at work in female sexual orientation than simply the levels of hormones in the womb.

With twin studies, psychological and mental testing, experiments on rats, and physical brain differences between homosexuals and heterosexuals, there is an overwhelming amount of circumstantial evidence that sexual orientation – at least male sexual orientation – is fixed or heavily influenced by hormone exposure in the womb. There is, as a district attorney would say, more than enough evidence to convict. But, the defense attorney protests, there is no solid evidence that my client, Mr Testosterone, pulled the trigger. Where's the smoking gun with his fingerprints on it?

The police haven't found it yet, but Detective Dean Hamer is getting very close. Hamer, a researcher at the National Institutes of Health, has apparently located a gene (or perhaps a set of genes) that produces a bias towards homosexuality in males. If his finding is replicated, not only will it be the first step toward understanding the biological basis for homosexuality, but it will rank as one of the great discoveries of the century – the first human behavior tracked down to its roots on the chromosomes.

To trace the homosexuality genes, Hamer recruited forty pairs of homosexual brothers and analyzed their genetic material, looking for a stretch of DNA on one of the forty-six human chromosomes that he could tie to the brothers' homosexuality. It works like this: if you have one pair of homosexual brothers, about half of their genes are the same and half are different; the genes for homosexuality should lie somewhere in the shared half. This doesn't narrow it down a lot – a human has about one hundred thousand genes, so with one set of brothers you cut down the candidates to only fifty thousand genes. But with a second set of homosexual brothers you can reduce it by half again, to twenty-five thousand, and so on. Theoretically, you would need less than twenty pairs of brothers to pinpoint the genes you're after.

It doesn't work this neatly in practice, for several reasons. It's possible, for instance, that different genes are responsible for homosexuality in different sets of brothers, or that the homosexuality is genetically caused in some of the brothers but triggered by something else – perhaps something in the childhood environment – in others. And scientists have not yet identified all of the genes a human has – only a very small percentage of them – so it's not really a matter of finding one or a few specific genes, but rather of zeroing in on a small span of DNA on one particular chromosome. If, however, Hamer could find such a span that is shared by a large percentage of the brothers, he could be pretty sure that the genes for homosexuality lie somewhere in that bit of DNA.

And he believes he has. Hamer started his search with the X chromosome because in earlier studies he had found that male homosexuality seems to be passed through the mother. If one male in a family is homosexual, then not only are his brothers more likely to be homosexual, but also his maternal uncles – his mother's brothers – and any cousins that are male sons of his mother's sisters. In some families, Hamer says, homosexuality does seem to pass through the paternal line, but it's not nearly as common as maternal transmission. And if male homosexuality is passed through the women in a family, Hamer notes, the genes for it should lie on the X chromosome. (A male, with one X and one Y chromosome, always gets the X from his mother and the Y from his father. Genes on the X chromosome are the only ones a son must have inherited from his mother, not his father.)

At one tip of the X chromosome, Hamer found a stretch of genetic material that was shared in thirty-three of the forty pairs of brothers. The odds are two hundred to one against this being merely a coincidence, Hamer says. If Hamer's results are reproduced by other scientists, then we will be able to say that a male's sexual orientation is determined, at least in part, by a gene or set of genes on his X chromosome.

No one can say exactly what this gene does. In fact, Hamer doesn't even know which gene it is. His technique involves the use of 'genetic markers', short pieces of DNA (the material that makes up the genes) that are used to identify particular sections of the chromosomes, so he knows which area of the X chromosome seems to be influencing sexual orientation but he doesn't know which or even how many genes are involved.

Still, Hamer can make a pretty good guess about what's going on – the gene likely plays some role in how male hormones organize the brain. It's probably not something as simple as telling the fetal gonads to make more or less testosterone because this would also affect such things as the growth of the male sexual organs, and these are, of course, completely normal in homosexuals. Instead, it's more likely that the gene affects how various parts of the brain respond to male hormones, perhaps in a way similar to what happens with XY women. As we saw in Chapter 3, these women are genetically male but become female because their male hormone receptors – the middlemen between the hormones and the body's cells – are defective and cannot detect testosterone or other male hormones. Although the brains of homosexual men can certainly detect male hormones, their receptors might respond somehow differently to testosterone than do the receptors in the brains of heterosexual men, or perhaps they might simply have a different distribution of receptors in their brains.

For now researchers can do no more than speculate about how a gene might determine sexual orientation, but soon they should know much more. With Hamer's identification of this one part of the X chromosome as playing a role in male homosexuality, it should be only a matter of time before he or other researchers locate the gene (or genes) exactly, determine its structure and figure out what it does. Once that is done, scientists will have a much better idea of what determines whether a man is attracted to women or to other men.

All of this applies only to males, Hamer notes. Although he has begun a study to look for genes involved with female homosexuality, no one knows whether female sexual orientation will be traced to the same part of the X chromosome – or even to the X chromosome at all. Hamer does say, however, that it's unlikely exactly the same genes are responsible for homosexuality in both males and females. In family studies, he has found that while fourteen percent of the brothers of male homosexuals were themselves homosexual (as compared with two percent of the general population by his rather strict definition of homosexuality), only five percent of their sisters were homosexuals. Similarly, only five percent of the brothers of lesbians were homosexuals. If the same homosexuality gene were involved in both sexes, he says, the brother–sister rates should be much higher.

So again, we find that we're further away from understanding female sexual orientation than male sexual orientation. Or, to put it another way, science is closer to explaining why (most) men find women attractive than why (most) women like men.

CHAPTER 7

Variations on a Theme

Looking around, you think you've wandered on to the set of a Grade-B science-fiction flick. In the middle of the room a man is lying on a table and placing his head into this mutant motor-cycle helmet which sprouts a dozen or so tubes connected to wires running off to hook up with a computer. Then, helmet on and breathing through a mask over his face, the subject closes his eyes and tries to relax. Over a pair of earphones a voice instructs him, 'You will hear a list of words. Each time you detect a word with exactly four letters, signal by raising your right and left index fingers slightly. Do not move any other part of your body...' Off to the side a mad scientist cackles and rubs his hands together with maniacal glee.

OK, so I made the last part up – there's no mad scientist. But the rest is true. It's a real-life scene from the laboratory of Cecile Naylor, a neuropsychologist at the Bowman Gray School of Medicine at Wake Forest University in Winston-Salem, North Carolina. Naylor, who is not mad and hardly ever cackles, uses the helmet/mask/earphones contraption to record which parts of a subject's brain kick into action when the subject performs various mental tasks. And in this simple, not-mad pursuit, Naylor stumbled across an extraordinary fact: men and women use different parts of their brains to perform the same job, even when the job is as simple as deciding whether a given word has four letters.

It all started, Naylor says, when she and her colleagues decided to peer into the brains of dyslexic patients – people of normal intelligence who have great difficulty learning to read – for a clue about what causes their reading problems. She wasn't looking for a sex difference at all. She figured that if dyslexics have some glitch in the way their brains work, it might show up as an abnormal pattern of brain activity recorded by her apparatus. So she recruited a number of volunteers willing to don a crazily wired motor-cycle helmet, breath through a mask and lie down on a table to take some simple tests while she monitored their brains.

The key test – asking which words from a spoken list have exactly four letters – is one that is very easy for normal readers but quite tough for dyslexics. For some reason the dyslexic brain is not well equipped to hear a word, visualize it, and count its letters, and Naylor suspected that if she monitored dyslexics' brains as they struggled with this task, she would find that the activity in their brains was different from that of normal brains doing the same job.

To observe that activity Naylor turned to her science fiction-y apparatus which, despite its appearance, is really quite simple. It works like this: when the subject is comfortably settled on the table with the helmet on his or her head, Naylor adds some radioactive xenon – a harmless, odorless gas – to the air mixture that the subject is breathing. After passing into the lungs, the xenon makes its way via the subject's blood to the brain. Once it is there, Naylor can trace it with the sixteen radiation detectors that are mounted on the inside of the helmet, each pointing at one carefully chosen spot of the brain. The harder an area of the brain works, the more blood flows through it and the more xenon the detector over that spot records. When the test is over, Naylor has a record of the activity of each of the sixteen parts of the brain.

If Naylor were going to recognize an abnormal pattern of brain activity in dyslexics, she first needed to know what a normal brain looks like in action, so she recruited a group of men and women with no reading problems to take the same test. Then, because her dyslexic subjects were mostly male, she broke the control subjects into males and females and analyzed them separately just to make sure that there weren't any unexpected differences between the sexes. There were.

In the normal subjects, Naylor says, one part of the brain 'lit up' more than any other while the subjects were counting the letters in words. It is a section of the brain just behind the left ear called Wernicke's area, and it is known to be indispensable to language and speech. In both males and females Naylor found that the more active Wernicke's area was, the more likely the subject was to get the correct answer on the letter-counting test. Furthermore, Wernicke's area was not nearly as active in the dyslexic brains as it was in the normal brains during the test.

So, Naylor concluded, Wernicke's area must play a major role in mentally spelling out a word, and dyslexia is somehow connected with a difficulty in activating this part of the brain in order to do certain tasks. Both male and female brains depend equally on Wernicke's area in spelling out words, Naylor found, but when she took the next step and examined which other parts of the brain were assisting, suddenly male and female brains looked like night and day.

Although Wernicke's area takes the lead in solving the letter-counting task, other sections of the brain also contribute, Naylor explains. Indeed, the whole brain lights up each time it tackles a problem, but some areas seem to be working closely together while others are relatively independent. So Naylor investigated which parts of the brain were co-operating most closely with Wernicke's area when the subjects tackled the simple spelling test. The answer, she discovered, depends on which sex is doing the spelling.

In males there was only one section that worked closely with Wernicke's area during the test – a region on the left side of the brain in front of the ear called Broca's area. The activity in these two areas, only a couple of inches apart in the left half of the brain, was tightly correlated, indicating that they were close partners in a collaboration to analyze the words the male subjects heard.

But the brains of the female subjects didn't act this way at all. When Naylor looked to see which areas of the female brain were working with Wernicke's area, Broca's area wasn't even in the picture. Instead, there were two other partners, one of them on the left side of the brain just behind Wernicke's area and the other one in the right hemisphere, directly across from Wernicke's area.

The wiring of the female brains was quite surprising, Naylor says, because it contradicts what most scientists believe about how the brain processes speech. The close co-operation between Wernicke's and Broca's areas that she saw in the male subjects is 'the classic language profile' of the brain, she says, but her results indicate that this picture of the brain may be true for only half the population. 'Most of the research in the past has been based on male brains,' she says, and scientists assumed that the same pattern would be true for female brains. Now, however, it seems that – at least when mentally spelling out words – men's and women's brains work in measurably different ways.

To date, Naylor's experiment is the most concrete and convincing demonstration of a difference in the way male and female brains are designed, but it's likely to be just the first in a long series. Only recently have researchers developed techniques that allow them to watch the brain in action, and so far these procedures remain complicated and esoteric enough that only a few researchers have applied them or even have access to them. But with more scientists learning to use the techniques and with more useful methods under development, the next few years should see a rush to observe the brain as it cogitates on various types of problems. As this book is written, for instance, Naylor is in the middle of an experiment using positron emission tomography (PET) to measure brain activity as subjects distinguish between letters of the alphabet and non-letter shapes. With PET, Naylor says, she will be able to get a much more detailed picture of the activity in the brain than was possible with the earlier experiments. She hopes eventually to get readings from some eighty to one hundred subjects, evenly divided between men and women – which should be more than enough to pin down any sex difference in the pattern of activity as the brain performs this very simple language test. The results, unfortunately, won't be available until after this book heads off to press.

Nevertheless, other scientists have already pieced together a pretty good idea of the differences in the way that men's and women's brains work. Even without the studies like Naylor's that should be appearing in the next few years, sex difference researchers have uncovered a number of details that distinguish the male brain from the female brain, and although it's still not clear exactly how all the pieces fit together, the big picture is slowly coming into focus.

That big picture consists of – as they say in music – variations on a theme. In this case the theme is the overall design of the human brain, which is approximately the same in males and females: the inner, 'primitive' part of the brain, which regulates instincts, drives and automatic body processes such as breathing and heartbeat; the limbic system, including the hippocampus, hypothalamus and amygdala, which serves as the emotional center of the brain; and the huge cerebrum and cerebral cortex, which give humans their capacity to think. Or more poetically, from inside to outside: heartbeat, heartstrings, heartless.

Working from this general theme, sex hormones act to create variations in the melody, making it more masculine or more feminine in myriad small ways but always leaving the tune recognizably human. Or, to muddy the musical metaphor, you might think of the 'masculine brain' and 'feminine brain' as point/counterpoint – two semi-independent melodies that often borrow phrases from one another and when played together can produce some beautiful harmonies. The bottom line is that the 'male brain' and the 'female brain' are two distinct variations of the human brain – although, as we shall see, these two versions of the human melody are actually more ideal than real. Your brain or mine will have a mixture of both masculine and feminine characteristics which combine to create a unique tune, neither fully male nor fully female.

Much of the current understanding of sex and the brain can be credited to two of the biggest stars of sex difference research. Each has been studying sex differences since the 1960s – longer than most, and long enough for each to have developed the expertise and perspective necessary to see the big picture. In a field dominated mostly by young Turks (young Turkettes?), these two from an earlier generation are the *grande dames* of sex difference research and they offer another version of 'variations on a theme'.

Doreen Kimura and Sandra Witelson got their start in the same place – McGill University in Montreal under the pioneering neuroscientist Donald Hebb, who was one of the first psychologists to study how behavior has its roots in the structure of the brain. Kimura was a graduate student under Hebb (and co-adviser Brenda Milner at the Montreal Neurological Institute) in the late 1950s, while Witelson came along to work with Hebb a few years later. Both have ended up at Canadian universities just down the road from one another – Kimura at the University of Western Ontario in London, and Witelson at McMaster University in Hamilton, less than a hundred miles away.

Despite their similar backgrounds and research interests, Kimura and Witelson have little in common, either in personality or approach to science. Doreen Kimura is driven, outspoken, often impersonal. One researcher I spoke with said he was honored that she considered him a colleague – implying that there were many other scientists to whom she does not accord that honor. Another researcher who likes and admires her speaks of her 'sharp tongue'. Certainly she speaks her mind, especially about her conviction that male and female brains are innately different. 'I would not expect,' she wrote in one article, 'that men and women would necessarily be equally represented in activities or professions that emphasize spatial or math skills, such as engineering or physics' – an opinion that bothers many people, including some fellow sex difference researchers. Kimura has written about her research for the public in such magazines as *Scientific American* and *Psychology Today*, but unfortunately for me, she doesn't make herself easily available to journalists who want to write about her work, and I was unable to convince her to agree to an interview for this book. (She was, however, very helpful and prompt in supplying any written material that I asked for, and she mailed reprints of articles and descriptions of her research within a day or two of getting my faxed requests.)

Less than seventy miles away, Sandra Witelson plays counterpoint to Kimura's point. Although Witelson appears just as driven as Kimura in her approach to research, her public persona is much less provocative. If asked directly, for instance, about the societal implications of sex difference research, her answers sound much like Kimura's, but she doesn't offer up those opinions in articles she prepares describing her work. She seldom writes for popular publications, but she does talk to journalists – if they can catch her. The only time she could fit me in during a several-week period was at the end of the day one Friday, but once she had finished her phone calls and paperwork she spoke openly with me for a couple of hours until she had to run off to get ready for the symphony. Like Kimura, Witelson is highly respected by her colleagues for her more than two decades of imaginative and careful research.

Kimura and Witelson have both established ongoing programs to collect brain data from hospital patients but, like the researchers themselves, the programs are quite different in purpose and tone. Whereas Witelson autopsies the brains of cancer victims, Kimura studies patients who have survived a stroke, brain tumor or other brain damage. The two approaches provide complementary looks at how male and female brains differ.

Since 1974, Kimura has been collecting data on brain-damaged patients at a neuropsychology unit she set up at University Hospital in London, Ontario. Her goal is to learn which parts of the brain govern which mental activities by studying how damage to various areas affects different brain functions. Over the years she has found that brain damage affects men and women in dissimilar ways.

Kimura's technique is not a new one. It's been in use for well over a century, dating back to Paul Broca, the French surgeon and brain researcher whom last we saw in Chapter 5 comparing the sizes of men's and women's brains. In 1861, by studying the autopsied brains of people who had lost the ability to speak at some point during their lives, Broca discovered that many of them had damage to one particular area – a spot about midway between the top of the ear and the outer edge of the eyebrow, now called Broca's area. He concluded, rightly, that this area plays a vital role in producing speech.

In principle you could use this technique to map out the brain quite thoroughly, but in practice it's not nearly so easy. Much of the brain damage that researchers have to work with is caused by strokes – a shut off of blood flow to parts of the brain – and stroke damage is seldom limited to one small portion of the brain. And even when the damage is more localized, which can be the case with tumors, it's often difficult to know exactly which parts of the brain have been affected and which haven't.

None the less, brain-damaged patients have taught doctors quite a lot about how the brain works, including some amazing lessons on just how specialized different parts of the brain can be. One group of doctors, for instance, reported a stroke patient who couldn't remember the names of fruits and vegetables. He was fine with people, places or things, but he couldn't come up with 'peach' or 'orange' even when shown pictures of them. Apparently one system in his brain was responsible for remembering names of fruits and vegetables but not other types of nouns. Another group of researchers reported evidence that nouns and verbs are handled by different parts of the brain. They found that some patients, shown a picture of a duck swimming in water, could say that it was a 'duck' but couldn't say what it was doing, while others knew it was 'swimming' but couldn't come up with its name.

I learned about such language loss firsthand in 1980 when my father was hit by a massive stroke. In the days after the stroke he couldn't speak at all, and for the next year or so he was regaining pieces of his language skills. He never got it all back, and our family learned to adapt to his deficits, which had a very definite pattern. No matter how hard he tried, for instance, he could never quite get the hang of pronouns: 'he' and 'she' were interchangeable to him, as were 'you' and 'I'. That made it quite confusing to hold a conversation about people – if he said 'he', I never could be sure if he meant my brother, my sister or my mother. Numbers were another sticking point. With enough concentration, he could count 'one, two, three...' but if he tried to say a single number, it was pure coincidence if he got it right. He'd look at the numeral '6' on a piece of paper and say, very clearly, 'three', repeating it several times and listening to himself as if he couldn't be sure what his tongue was saying.

No researcher ever recruited my father for a study, but plenty of stroke victims like him have helped doctors learn how the human brain is organized. The clearest finding is that the brain carries out very different types of functions in its right and left hemispheres.

Language skills – listening and speaking, reading and writing – are normally concentrated on the left side of the brain, as is the higher-level control of hand and arm movements. (Like many other left-side stroke victims, my father's difficulty with language was accompanied by a general clumsiness. The man who just the year before had built a beautiful, incredibly detailed model of a sailing ship now had trouble picking up a piece of food with a fork.) By contrast, the right side of the brain has primary responsibility for a number of non-language functions, such as recognizing faces, responding to emotions – including the emotion expressed in speech – and comprehending spatial relationships, such as reading a map or a blueprint.

This is the general pattern, but researchers first began to notice in the 1960s that it doesn't work exactly the same way for men and women. Among people who suffer a stroke or other damage to the left side of the brain, men are much more likely than women to suffer speech disorders – aphasia, to use the technical term. If it had been my mother instead of my father who was hit by a stroke, she would have had a better chance of keeping her language skills. Such sex differences, brain researchers concluded, must surely point to some variation in how men's and women's brains are arranged, and for the past fifteen years they have been trying to sort it out.

In the late 1970s, for instance, Jeannette McGlone, a psychologist at University Hospital in London, Ontario, offered a simple hypothesis: women's brains are more symmetrically organized than men's brains. McGlone, who had received her Ph.D. under Kimura, suggested that both speech and non-speech functions are spread out more evenly over the two hemispheres in women. Men's brains were more asymmetric, she suggested, with speech controlled mostly on the left side and the right side reserved for non-verbal functions such as spatial ability and recognizing patterns.

It was a beguiling idea. Not only would it explain why women are less likely to suffer aphasia after damage to the left hemisphere – they can fall back on the right hemisphere, while men can't – but it might also account for men's advantage in spatial ability and women's superiority in verbal skills. Indeed, University of Chicago psychologist Jerre Levy had already speculated that this might be the case. According to her 'cognitive crowding hypothesis' described in Chapter 5, since language is so important to humans, spatial skills would suffer whenever one part of the brain attempted to handle both them and verbal skills. Levy suggested that the right side of men's brains may be specialized for spatial skills and the left for verbal skills, giving men better spatial ability because spatial thoughts are not inhibited by verbal processing. Women, with both hemispheres working together on language, would have an advantage in verbal skills over men, who have only one verbal hemisphere.

The theory had the extra advantage that it was easy to explain to the general public: men's brains are more compartmentalized; women's brains are more holistic. With all that going for it, it's a real shame that the theory just wasn't quite right.

The cracks started to appear after 1980, once people looked more closely at how strokes in different parts of the brain affect men and women, McGlone says now. If the right side of the male brain were more specialized for spatial ability and other non-verbal functions, then damage to the right hemisphere should hurt these non-verbal skills in men more than in women. But several studies have shown that doesn't happen, McGlone says. On the other hand, if women's speech centers are more spread out and the right hemispheres play a greater role in language, then damage to the right side should impair women's verbal skills more than men's since men have little verbal processing going on in the right hemisphere anyway. Again that's not the case, she says.

But what else could explain the fact that women are less likely to lose their language skills after a stroke or other damage to the left hemisphere? Kimura, working with a database that has grown to include hundreds of patients, believes she has an answer.

Not all left-side injuries are the same, Kimura says. Some patients have damage spread over the entire hemisphere, but in others it may be localized in just one section. In one study of two hundred and sixteen patients whose left hemispheres had been damaged by stroke, tumor or other disorder, Kimura found eighty-one in whom the damage was restricted completely to either the front half or back half of the hemisphere. And in these patients she found her clue to the sex difference.

The females in this restricted-damage group were more likely to suffer aphasia, or language problems, if the damage was in the front half, Kimura discovered. This implied that the critical speech areas for women were in the front part of the left hemisphere. The males, on the other hand, were more sensitive to damage in the back half of the hemisphere, so more of their speech must be controlled in the rear. By itself, this difference wouldn't explain why women are less vulnerable to strokes on the left side, but Kimura also found that strokes hit the back half of the brain more often than the front – in both women and men.

This was the key, she said: if strokes strike the rear half of the left hemisphere more often than the front half, and if men have more of their critical language areas located in the back half than women, then men will more often lose their language skills after a left-hemisphere attack. It's not a matter of women's language centers being more spread out than men's, but simply that the parts of a woman's brain that control language are in a safer place.

Since Kimura published those findings in 1983, she has continued to collect information on the differing effects of brain damage to men and women, and she now has data that turn the original hypothesis about the sex differences in the brain upside down. It is men, not women, she says, whose language centers are spread out over more of the brain and it is women who have the critical language areas concentrated in a smaller area.

To divine this, Kimura divided the left hemisphere into five sections – anterior, central, temporal, parietal and occipital (going from front to back) – and considered how damage to each of these areas was likely to affect men and women. It takes many patients to do this, Kimura notes, since strokes tend to be spread out over much of a hemisphere. Among her subjects, less than half the victims of left-hemisphere strokes had damage that was restricted to a small area. But when she analyzed the minority with limited damage, she found that men's and women's brains react very differently to a stroke in a particular spot of the left hemisphere.

Women are most vulnerable to harm in the anterior, or front, region, she says. Among the female patients who had strokes or other damage there, about two-thirds lost some language skills. But damage to the rest of the hemisphere hardly affected language in women at all. None of the women with injury to the central, parietal and occipital regions lost their ability to speak and understand language, and only a few of those hit in the temporal region developed aphasia.

Men, on the other hand, were vulnerable to language loss if a stroke hit any of the five regions of the left hemisphere. They were most susceptible in the parietal region – a section that seems to have few language functions in women – with damage to the parietal causing language loss in fifty percent of the male patients.

Kimura concludes that the language regions in men's and women's brains are situated in very different places. Women's language skills seem to be concentrated mostly in the front part of the left hemisphere, while men's skills are widely distributed across most of the left side, with more of them in the back half.

A similar thing is true for the regions of men's and women's brains that direct the movements of the limbs, Kimura says. She tested the 'motor programming' of her brain-damaged patients by asking them to imitate a simple series of hand and arm motions – blowing a kiss, for instance, or perhaps first touching the back of one hand to the opposite forearm and then turning the hand to tap the palm to the forearm. Although most people have no trouble with such movements, some of Kimura's patients just couldn't make their bodies co-operate. Instead of a fingers to the lips, it might be a fist to the nose or a palm to the chin.

Almost all of the patients with this 'apraxia' – difficulty in directing the arm or hand to do a series of movements – had damage to the left hemisphere. This was true for both males and females. But the sexes differed dramatically when Kimura looked at damage to the front part of the left hemisphere versus the back part. Among the women, seventy-one percent of those with injury to the front section had apraxia, but only seven percent of those with damage in the back did. In men the pattern was reversed: only twelve percent with frontal damage had difficulty moving their hands through complicated series of motions, but forty-four percent of those with injury to the rear section did. In women, Kimura concluded, the control of complicated hand movements seems to be concentrated in a smaller area in the front of the left hemisphere, while in men it is more spread out with a greater percentage of control toward the back.

These differences in brain organization may have practical, noticeable effects in men's and women's abilities, Kimura speculates. In women the critical areas for organizing hand movements appear to be in the front part of the left hemisphere, which puts them close to the motor cortex, the section of the brain that sends out signals telling the body's muscles what to do. Perhaps, Kimura says, this proximity may explain why women have better control of small, precise hand movements. Men, on the other hand, excel at muscle movements directed at distant objects, such as throwing a baseball or firing a gun at a moving target. Since these actions demand co-ordinating muscle control with sight, it makes sense that males have more of the control of their limb movements located toward the back of the hemisphere, closer to the visual cortex.

Kimura's work on the brain-damaged patients comprises the most extensive analysis of sex differences in brain function that anyone has done – or is likely to do any time soon. She has worked for nearly two decades to collect her data on hundreds of people with strokes, tumors and other sorts of brain damage, and it would take years for another researcher to come up with comparable information. Still, there are a few other experiments against which Kimura's results can be checked.

In 1989, for instance, Richard Lewis and Lois Christiansen at Pomona College came up with a way to test Kimura's ideas on healthy people. Suppose, as Kimura says, that women's brains concentrate both speaking and the control of muscle sequences in the front half of the left hemisphere, while those two functions are more spread out over the entire left side in men's brains. Then if you ask a person to do both things at once – speak and move a hand or arm in some pattern – it should be harder for a woman than a man. The two actions should interfere with each other more in a woman's brain since one part of the brain will be trying to do two things at once.

To test this idea, Lewis and Christiansen recruited fourteen male and fourteen female undergraduates. First they had each of them perform a finger-tapping test. The subjects used the four fingers on one hand to push down four buttons in sequence – index finger, middle finger, ring finger and little finger – again and again, as many times as they could in thirty seconds. The subjects were scored on how many sequences they completed with each hand. Then the researchers had them do it again, but at the same time the subjects had to either read a passage from a book or name as many words as they could think of that started with a particular letter. Afterward, Lewis and Christiansen compared the tapping scores with and without the interference from the language tasks.

Both males and females slowed down when they had to read or list words, but the patterns of the slow-downs were very different. For the men it didn't matter much whether they were tapping with the right or left hand – the hold-up was about the same. Among the women, the slowing of the left hand was about the same as it was for men, but right-hand tapping slowed much more. Since muscles on the right side of the body are controlled by the left half of the brain, the subjects were performing as Kimura would predict – in women, the simultaneous language task and right-hand tapping were interfering with each other more, presumably because the control of both skills is centered in a much smaller area of women's left hemispheres than of men's.

In 1982, a researcher who had earned her Ph.D. under Kimura tested Kimura's ideas in a completely different way – by temporarily knocking out individual parts of human brains and seeing what happened. Catherine Mateer, collaborating with Samuel Polen and George Ojemann, looked for sex differences in how the brain handles language by taking advantage of a curious feature of a type of brain surgery performed on epileptics.

For some patients with severe epilepsy that can't be controlled with drugs, the only treatment is to remove a part of the brain where the epileptic attacks begin. Surprisingly, people can lead normal lives with relatively big chunks of their brains missing, so the surgery can be effective as long as the doctors are careful not to cut out anything vital. In particular, when the doctors operate on the left hemisphere, they must be careful to avoid the areas that are necessary for speech, and the only way to be sure which areas those are is to test for them in each individual patient.

That's where Mateer and her co-workers came in. Ten female and eight male patients were scheduled to have parts of their left hemispheres removed to stop epileptic seizures. At the beginning of each surgery, after the doctors had opened up the brain but while the patient was still awake, the researchers gave the patient short tests of language and memory while at the same time running a mild electric current through different parts of the patient's brain. (The brain has no pain sensors, so the test was not uncomfortable for the patients.) In a 'naming test', for instance, the researchers showed the patient slides that each contained a short phrase such as 'This is a . . .' plus a drawing of some common object, and the patient had to name the object. If he or she couldn't come up with the name of the object while current was being applied to some part of the brain, the doctors assumed they should avoid that particular area.

When Mateer, Polen and Ojemann analyzed which sections of the brains were critical to the naming task, their basic conclusion supported Kimura's work: the speech areas were much more concentrated in the women's brains than in the men's. In the eight male patients, the electric current disrupted speech in sixty-three percent of the sites tested, but among the female patients only thirty-eight percent of the sites were critical for the naming test.

The researchers also agreed with Kimura on the area with the greatest difference between men and women – the parietal, or upper rear, area of the left hemisphere. Among the males, about two-thirds of the sites tested in the parietal area were affected by electric current during the naming test; for the females, the number was less than twenty percent.

The results didn't completely jibe with Kimura's, however. Mateer and company found no evidence that the front part of the left hemisphere is more important for speech in women than in men. In one zone in the front part of the brain, for instance, electric current caused all the patients to make mistakes on the naming test, both males and females. In another frontal portion the males were actually more sensitive to the current than the females. The researchers had no explanation for why their results disagreed with Kimura's, but the discrepancy may be related to the sizes of the regions tested – the electrical stimulation during surgery affected very restricted pieces of the brain, while Kimura studies the effects of strokes that hit much larger areas.

Such seeming inconsistencies are relatively common in brain research today. No one can say, for instance, how Naylor's finding that women use part of their right hemispheres in language processing fits in with Kimura's results on the women's concentration of language skills in the front of the left hemisphere. And if you stop to think about it that's not too surprising – the two researchers may be working on the same puzzle, but they're putting pieces together in totally different areas. Eventually, when we know more about the human brain, we should be able to see how Naylor's section of the puzzle connects with Kimura's, but for now we have to be content with contemplating the isolated details.

Thanks in large part to Kimura, we have a great many of those details to contemplate about the left side of the brain – in particular, how language and muscle control are handled by different parts of that hemisphere in men and women – but the right side is much less explored. Judging from a few studies, Kimura suspects that the picture in the right hemisphere is similar to that in the left, with the division of labor between the front and the back different in men and women. For instance, when Kimura tested men and women with damage to the right sides of their brains by asking them to put together blocks to form given patterns, she found that the women scored worse if they had damage to the front of the hemisphere and the men worse if they had damage to the back. The results are still preliminary, though, and need to be replicated and extended.

There may also be sex differences, Kimura thinks, in the way the brain splits up duties between its left and right sides – although it's not as simple as once thought, with women's brains being more symmetric and men's brains more specialized. Certainly Cecile Naylor's research points in that direction, with her finding that in spelling out words, men's brains relied mostly on two pieces of the left hemisphere, Wernicke's area and Broca's area, while women's brains used Wernicke's and two different sections, including one on the right side of the brain.

Kimura hypothesizes that differences in right–left organization may be important in more abstract mental processes, even though they don't seem to appear in such abilities as basic speaking skills or co-ordination of hand and arm motions. She finds, for instance, that on tests of vocabulary, women's performance seems to be impaired by damage to either the left or right hemisphere, while men's scores are worse only if the damage is on the left.

And Richard Lewis at Pomona, this time working with Laura Kamptner, has uncovered evidence that spatial ability may be more focused in the right hemisphere in men than in women. Like Kimura, these researchers studied men and women whose brains had been damaged on one side or the other. They gave the subjects two tests – a 'completion' test, in which a person must figure out what a picture looks like when pieces of it are missing, and a 'block design' test, which requires a subject to assemble a number of cubes, each with certain geometric patterns on its sides, so that the patterns create a given design – and then compared the patients' scores with those of a group of people without brain damage.

On the completion test, which doesn't normally show a sex difference, they found that men and women were affected similarly by damage to the right or left hemisphere. Left-side harm affected both sexes' performance somewhat, and right-side harm hurt it much more. That was expected, since spatial ability is mostly concentrated on the right side.

But on the block design, a test on which men usually score higher than women, there was a dramatic difference. The men didn't seem affected at all by damage to the left hemisphere, but those with injury to the right hemispheres posted scores little more than half what normal men get. Women, on the other hand were almost equally affected by damage to left or right side. Lewis and Kamptner's conclusion: the type of spatial ability measured by the block design test – a skill that men excel on – is focused in the right hemisphere in men but is more spread out over both hemispheres in women. The same is not necessarily true, they said, for other types of spatial ability.

So, fifteen years after Jeannette McGlone suggested that the sex difference in responding to strokes could be explained if women's brains were more symmetric than men's, the symmetry issue remains confused. It does appear that for some functions women's brains are more likely to involve both hemispheres while men's brains rely on just one side. But in other cases researchers see no difference in how the sexes divide work between the hemispheres. None the less, in the bewildering mass of data that has accumulated over the years, one fact stands out: whenever a study does find a sex difference, it always points in the same direction – women's brains are more symmetric. Few, if any, experiments have found a mental function that is handled by both hemispheres in men but only one in women.

No, this doesn't mean that women's brains are more balanced (as my wife claims), but it does seem that being a female makes it more likely your brain will distribute some of its functions over both the right and left sides.

Sandra Witelson has taken a road quite different from Kimura's. Indeed, she insists that she has never been interested primarily in sex differences. Instead she tries to understand how behavior is linked to brain structure. Are there, for instance, anatomical differences between the brains of right-handers and left-handers? But inevitably sex gets mixed up in her studies.

As a graduate student Witelson looked at learning problems in kids with brain damage, and in the early years of her career she focused on children with dyslexia. To understand why these children of normal intelligence had such difficulty learning to read, she studied the interplay between their right and left hemispheres using 'dichotic listening' – a type of experiment in which the subject hears one thing in one ear and simultaneously something else in the other ear and is tested on what he heard. In general one ear will be dominant – the brain pays more attention to what it hears from that ear than from the other – and this reflects a dominance by one hemisphere or the other.

From hearing, Witelson moved on to touch, and here she found her first sex difference. Her goal was still to see how the two halves of the brain compared, and she came up with an imaginative test. A young subject reaches behind a screen and is handed two objects with unusual shapes, one in each hand. The subject has ten seconds to manipulate the objects and then is asked to pick out the two shapes from a set of six drawings. Because information about the object in the left hand is received by the right hemisphere and vice versa, and since the right hemisphere is believed to process information about shapes, Witelson thought that the subjects might be more accurate in identifying objects held in the left hand.

But when she tested two hundred boys and girls at ages six through thirteen, she found that this was only true for boys. The girls did just as well with objects in one hand as the other, while the boys were better with objects in the left hand. Although all the children got better at identifying the objects as they got older, the left-hand advantage appeared in the boys at all ages and in the girls never. In males only, Witelson concluded, the right hemisphere does more of the work in identifying shapes – a type of spatial skill. (Another possibility is that the right hemisphere does this work in females too, but that the two hemispheres of a female brain communicate with each other better, so the right hemisphere gets just as much information about objects in the right hand as the left.)

After that 1976 study, Witelson says, she began to wonder if she could find a structural difference in male and female brains that was behind this mental difference. Perhaps by comparing the right and left halves of male and female brains she could find something that explained why males had spatial ability concentrated on one side.

So she began a project that has now continued for more than fifteen years and which is surpassed only by Kimura's studies of brain-damaged patients as the longest-running collection of data on sex differences in human brains. Witelson works with cancer patients whose tumors have metastasized and whose prognosis is not good, but whose cancers do not affect the brain. While they are still in relatively good health, she runs them through a large battery of tests which focus mostly on cognitive abilities and brain lateralization – which side of the brain is dominant for various factors, including such things as handedness and dichotic listening. They also agree to have clinical autopsies after they die, which will include careful analyses of their brains' anatomies. (The next of kin must give final permission for an autopsy after the patient dies.) In this way Witelson can look for connections between brain structure and function.

Although at first it may seem an almost ghoulish practice – waiting for cancer patients to die in order to examine their brains – Witelson points out that the patients are going to die one way or the other, and they themselves are usually quite eager to take part in the experiment, recognizing that they can make a difference by simply taking a few tests.

'The subjects get something out of it. They feel they're making a long-lasting contribution to mankind.' Since researchers normally know very little about the people whose brains they examine at autopsy, Witelson's study offers a rare opportunity to investigate the brains of people whose mental characteristics are known in advance.

Witelson took advantage of this opportunity in the mid 1980s to look into the controversy that had grown up around the corpus callosum. After Marie-Christine de Lacoste announced in 1982 that she found the splenium – the rearmost twenty percent of the corpus callosum – to be larger and more bulbous in women than in men, a number of researchers had tried to reproduce her discovery. When Witelson looked at the corpora callosa from her cancer patients, she saw no sex difference in size, but she did spot one surprising detail that others had missed: the corpus callosum was larger in left-handers and ambidextrous people than in right-handers. It was a discovery that could not have been made without examining the subjects before they died, since autopsy reports don't include handedness data and even questioning the family afterwards is often unreliable. (Now scientists are increasingly using magnetic resonance imaging to study patients' brains while they are alive and healthy, but at the time autopsies were the only option.)

As Witelson continued to collect brains and study the callosa, she stumbled across another, even more unexpected item: the difference in the size of the corpus callosum between right- and left-handers was true only for men. The callosum was the same size in right-handed women as it was in the group of non-right-handers, which included both left-handers and ambidextrous women. But among the men, the corpus callosum was an average of about fifteen percent larger in the non-right-handed group.

Why should handedness have anything to do with the size of the corpus callosum? And why should the correlation appear in male but not female brains? Witelson thinks the answer lies in the way that sex hormones influence the development of the brain.

Although researchers still don't understand why some people are right-handed and some are left-handed, they do know that handedness reflects a person's underlying brain organization and which side of the brain is 'dominant' in controlling muscle movement. And there is some connection between sex and handedness since there are more males than females who are left-handed.

Furthermore, handedness involves more than simply which hand you write or brush your teeth with or which foot you prefer in soccer – it's a mental characteristic as well. Almost all right-handers have language concentrated in the left hemisphere, for instance, while only about seventy percent of left-handers do, with the remaining thirty percent having language control in the right hemisphere or spread over both sides of the brain. (This leads many researchers, such as Kimura with her work on brain-damaged patients, to look only at right-handed subjects, and many reports of sex differences apply only to right-handers.) And for reasons that no one understands, left-handers are over-represented among the people with the highest IQ scores but are also more likely than right-handers to have severe reading disabilities.

Witelson believes that being right- or left-handed implies different things for men and women. Right-handed men's brains, she suggests, are 'lateralized' – they split up duties relatively strictly between right and left sides. Left-handed men, by contrast, have a different brain organization that spreads things out more evenly between the halves. And women's brains are more symmetric than those of right-handed males, no matter whether the woman is right- or left-handed. If so, this would explain Witelson's findings on the corpus callosum – the callosum in right-handed men is smaller because their hemispheres don't talk to each other as much.

Because more males than females are left-handed, most researchers have assumed that handedness is somehow tied up with the level of testosterone in the womb. The late Norman Geschwind, in an influential 1982 paper, suggested that high levels of testosterone slow the growth of the left hemisphere in the womb, causing the right hemisphere to become dominant and making a person left-handed. Witelson, on the other hand, suspects that it is the other way around – in males at least, it is low levels of testosterone that make a person more likely to be left-handed. She doesn't have a suggestion for what makes a female left-handed, but she's in good company. Nobody seems to have any ideas.

No matter. Like the view through a steamed-up shower door, the general outline can be made out even if the details must still be left to the imagination. Witelson offers the following sketch of what goes on in the brain during development, and most sex difference researchers would agree with the pattern if not the specifics.

Like the body, the basic design of the brain seems to be female in the sense that if the developing brain is not exposed to male hormones it will naturally move down a female path. Keep in mind, however, that this female path does not have just one destination – as with bodies, there is a great deal of variation from one brain to the next that has nothing to do with sex hormones. Two women exposed to the same levels of estrogen and testosterone in the womb can have quite different IQs, verbal abilities, spatial skills and psychological traits, such as nurturance or aggressiveness. Human brains naturally differ from one to another.

On top of that normal variation, testosterone creates another level of difference – it pushes brains in a masculine direction. Testosterone in the womb produces rough-and-tumble play in children, makes kids prefer trucks to dolls, increases aggressiveness later in life, improves spatial ability and probably leads to people who would rather look at maps than scenery. It may slow down the maturation of the brain in humans as it does in rhesus monkeys. It probably influences a person's sexual orientation, so that too little testosterone in a male or too much in a female may create sexual attraction to the same sex. It may affect just where in the brain such functions as language and spatial reasoning are handled, and it likely plays some role in determining handedness, although that's still up in the air. Again, keep in mind that these hormone-created changes come on top of differences that already exist in the brain. If one man has more spatial ability than a second man, for instance, it doesn't necessarily mean that the first got more testosterone in the womb.

Compared to testosterone's very active part in shaping the brain, the role of female hormones is not nearly so well understood. Some researchers, such as Dominique Toran-Allerand of Columbia University, argue that the basic human brain is actually 'neuter' and that a certain minimum level of estrogen is needed to create a female brain – in other words, that feminization is an active process like masculinization is. In rat brain tissue, for instance, Toran-Allerand has shown that estrogen stimulates certain types of neuron growth. And the fact that Turner's women, who have abnormally low levels of estrogen, have problems with memory and word fluency indicates that estrogen may be necessary to the development of these abilities.

Indeed, estrogen may be essential to life. Testosterone certainly isn't. After all, there are people alive whose bodies cannot detect male hormones – the XY women, for instance, who are genetically male but appear female because their bodies do not respond to any male hormones. Except for having no ovaries or uterus, however, they are completely normal. But, Toran-Allerand notes, there are no known cases of a person whose body cannot detect and respond to estrogen. The implication is that if estrogen can't do its job, you're dead. You're never even born.

So the question of whether the basic human brain is female or neuter may be more a matter of semantics than anything else. Toran-Allerand sees estrogen as feminizing a neuter brain. Other researchers tend to think of estrogen's role as essential for creating a normal human brain, which is female unless exposed to male hormones. But one way or the other, the picture is the same: exposing the brain to male hormones is like flipping a switch. Without them, a normal brain develops along a female path, but throw in some testosterone and the brain takes off in a masculine direction.

This does not mean, Witelson says, that a brain is either all masculine or all feminine. Unlike the sexual organs which develop during one short period in the womb, the brain develops over many months, during which time the levels of sex hormones can change dramatically. Since different parts of the brain are vulnerable to the hormones at different critical periods, one area of the brain can be masculinized while others aren't. Further, it seems that the level of testosterone is probably not the same all across the brain, so different areas get different doses. And finally, depending on the individual, various parts of the brain may respond more or less strongly to the presence of sex hormones. One area of the brain might be greatly masculinized while another remains relatively untouched.

Put it all together and you have what Witelson calls a 'neural sexual mosaic' – a patchwork brain whose pieces are each masculinized (and perhaps feminized) to different degrees. It's as if you handed a three-year-old a couple of crayons and a coloring book with a drawing of a brain in it and said, 'Make a pretty picture'. Now if you'd given the job to a seven-year-old you might get a brain that's all pink or all blue with perfectly even shading throughout, but three-year-olds don't do monochromatic marking. A little blue here, a lot of pink there, all thrown together with a grand disdain for neatness and order – that's your brain.

Because male fetuses have relatively high levels of testosterone, their brains are colored mostly blue, but pink patches can slip in here and there. Females, with much less testosterone circulating during development, generally stay on the pink side, but there's nothing preventing a little or even a lot of blue. (And keep in mind that there is plenty of natural variation in brains independent of hormones – each of the brains that the three-year-old is coloring is drawn differently from the others.)

This idea of the brain as a 'neural sexual mosaic' fits perfectly with what we know about people. Men and women are blends of masculine and feminine traits, a little of this, a lot of that. As a group, women have better verbal skills and better dexterity, are more nurturing and more interested in people, and look for solutions to problems that take everyone's interests into account, and as a group, men have better spatial ability, are more aggressive, competitive and prone to violence, are interested in objects and facts more than people, and tend to think of solutions more in terms of right and wrong. But to see men and women as members of these two mutually exclusive groups is to create an artificial distinction, one that is true on average but not necessarily true for any given man or woman.

Throughout the researching and writing of this book I've constantly compared what I've learned from scientists with what I know about the man and the woman I understand best – me and my wife. At times we fit the stereotypes so well I just have to laugh, in other cases we are exceptions, and when it is all added up I can see that we are each a mosaic of masculine and feminine.

In most ways I'm a stereotypical male. I'm good at math, I have high spatial ability and I love maps, blueprints and diagrams of any sort. My favorite thing to do is to learn how things work. On the other hand, I'm not physically aggressive and as a boy I avoided fights when I could. My wife tells me I'm intensely competitive. Perhaps my most feminine trait is that I love language and words, especially reading and writing. In talking, however, I'm a little slow and often I can't find the right word. I don't make emotional contact easily with other people and my wife says I have absolutely no empathy, which is probably true.

By contrast, Amy is a blend of the very feminine and the very masculine. She is intensely nurturing and has made friends with every child on the block. They all know they're welcome at Amy Pool's house. She is compassionate, understands people instinctively, and places great value on having a web of friends and family around her. She's also extremely articulate and can outtalk me by two or three to one. On the masculine side, Amy is very aggressive, and in professional settings this is usually the first impression one gets of her. She judges actions, both hers and those of others, on a strict standard of right and wrong and is not shy about telling anyone – family members, friends, teachers, co-workers, bosses and especially me – when their actions are not fair or just. Her ideal career would be to work for the FBI on anti-terrorism or in their psychological unit (she wanted to do this long before Jodie Foster made it famous in *Silence of the Lambs*), but she's decided to work in a less dangerous area of law and justice since she needs a career she can combine with raising a family.

Our next-door neighbor has dubbed Amy 'Donna Reed with a twist'. In our house, the running joke is that Amy is a 'he/she' because we don't really have a good word to describe someone who combines so many masculine and feminine features in one body. The technical term is 'androgynous', which the dictionary defines as 'having both male and female traits', but we've never liked that word because it has come to mean someone in whom the dividing line between male and female is blurred, a person you look at and can't be quite sure if it's man or woman. That's not Amy. No one has ever doubted she's a woman. They're just surprised to find so many traits they usually expect in a man packed into a female body.

The surprise seems to stem from the generally accepted idea that 'masculine' is the opposite of 'feminine', but the more researchers learn about the sexes, the less true this seems. When psychologists test people on how masculine or feminine they are, they actually use two scales, a masculinity scale that measures things like physical aggressiveness, dominance and risk-taking, and a femininity scale that measures nurturance, emotional responsiveness and other such characteristics. A person's scores on the two tests are relatively independent of each other – a stereotypical male would score high on masculinity and low on femininity, for instance, but many people score either high or low on both scales. High on both scales puts you in the androgynous category, while low on both makes you 'undifferentiated'. Instead of thinking of masculine and feminine as opposing categories so that a person is either one or the other, psychologists think of them as independent, almost complementary traits.

The same is true for other characteristics that show sex differences. Paying attention to landmarks is not the opposite of using a sense of direction – they're simply two approaches to the same problem, and if a person can combine both of them, so much the better. Spatial ability may be a masculine trait in the sense that males score higher on mental rotations tests, but you certainly wouldn't say that having poor spatial ability makes a woman more feminine. Females have better verbal skills, but being a good writer or speaker doesn't make a man less masculine. All of these traits can coexist peacefully in one person.

In this sense, I've come to see Amy as the most *human* person I've ever met. Her attributes – nurturance, aggressiveness, compassion, competitiveness and more – are first and foremost human traits, and she has them all in rare abundance.

There is an important lesson here, I think. It's easy and convenient to think of humanity as split into two sexes, male and female, and in a physical sense that's true. Women can bear and nurse children, men can impregnate women. But sometimes the physical differences combined with the average differences in mental and psychological traits can trick us into seeing the sexes as more distinct than they really are. There is far too much mixing and intermingling of male and female traits in people for 'masculine' and 'feminine' to be more than weak approximations of a person's mind or brain. Is an aggressive, nurturing person masculine or feminine? What about someone with high spatial skills and equally high verbal ability? And how would you label, say, a female homosexual who is not aggressive, has good verbal skills and doesn't like children?

All of these variations and many more are possible under the hormonal system that colors the development of the human brain in the womb. It is a complex, intricate, unpredictable system which guarantees that we'll never know exactly what to expect when we hear the words 'It's a boy' or 'It's a girl'.

CHAPTER 8

Raging Hormones

To a psychologist, college students are the equivalent of white rats. They're cheap (relatively), easy to work with, and there is a virtually inexhaustible supply. Walk into the psychology building on any university campus and you'll see notices hung on the walls and especially in the elevators – where people are trapped with nothing else to read – soliciting students for experiments. 'Wanted: Twenty right-handed and twenty left-handed subjects for a study of language skills', or 'Wanted: Fifteen subjects for an investigation into the effects of large amounts of alcohol on co-ordination, memory and sex drive. Fraternity members need not apply'. The pay for taking part in the experiments, which usually last an hour or two, is pretty poor – often no more than ten or twenty dollars – but college students always seem to need the money.

And so it was that in the spring of 1992, when University of Western Ontario psychologist Elizabeth Hampson and her student Christine Szekely had an idea for a new experiment, they put up a notice. It read something like, 'Wanted: Female subjects taking birth control pills for a study on how hormones influence performance on psychological tests'. They offered 'the going rate' for psychology experiments, Hampson remembers, and pulled in about sixty volunteers, most of them students at the university.

It was an easy ten bucks. Each subject filled out a short questionnaire and took a few paper-and-pencil tests. One of them was the Vandenberg mental rotations test, on which the testee must decide whether one three-dimensional figure is the same as a second figure drawn from a different angle. To get the answer, a person must visualize the figures and mentally rotate them to see if they match up, but they have to do it fast –

the subjects were given eight minutes to answer twenty questions with four rotations apiece.

The Vandenberg test is a standard measure of spatial ability, and it is a favorite of sex difference researchers because it consistently shows a large sex difference, with males scoring better than females. In statistical terms the average male score on the test is usually from three-quarters to one full standard deviation higher than the average female score. This is a big difference: it means that only one of every six females scores better than the average male, and five of every six males outperforms the average female.

Hampson and Szekely wanted to see if and how the hormones in the Pill would affect women's spatial ability, so they gave the test to half of the subjects during the menstrual week of their cycles, when they were taking no birth control pills, and to the other half sometime during the remaining three weeks, when they were taking the pills. Because birth control pills contain a synthetic hormone similar to estrogen, the second set of subjects had much higher levels of female hormones in their bloodstreams when they took the test than the menstrual group did. That was the only difference between the two groups.

But when Hampson and Szekely compared the scores of the two groups on the mental rotations test, they found a major disparity: the women who had taken the test during their menstrual period scored much higher on average than the other half of the subjects. The difference was about half a standard deviation, or nearly two-thirds of the average gap between males and females on this test.

What happened? Not only is this a surprisingly big difference, but it seems to be pointed in the wrong direction. If the menstrual group had scored *lower*, you could blame it on the discomfort that many women report during the week of their period, but those women had come out ahead.

There seems to be only one explanation, Hampson says: The female hormones in the Pill apparently do something to the brain that hinders spatial ability. Women do mental rotations faster and more accurately when they're not exposed to high levels of those hormones.

It's an uncomfortable idea, Hampson admits. Tens of millions of women take the Pill every day – how could something this common, this much a part of everyday life, be playing games with women's minds and nobody ever noticed? Part of the answer, she says, is that the effects of the hormones in birth control pills really aren't that large. Yes, a woman on the Pill might score lower on the Vandenberg test during one part of her cycle, but she could still do the mental rotations. All of the women tested were still well within the normal range, Hampson notes, with some of them scoring higher than many men. A woman would just be a bit slower, make a few more mistakes when taking the pills than not. And after all, Hampson notes, visualizing a three-dimensional object and mentally rotating it is not something that most women – or most men, for that matter – do every day. It's a limited, very specialized skill.

But there's a second, more important part of the explanation for why women might not notice such an effect: the synthetic hormones in the Pill aren't really doing anything unnatural to the brain. They're simply doing the same thing that a woman's natural hormones do as they rise and fall throughout the menstrual cycle. Hampson and other researchers have shown that the same sex hormones that produce a male or female in the womb continue to work on the brain decades later, fiddling with the settings of the masculinity/femininity knob.

Throughout a person's life, researchers have discovered, sex hormones act to 'fine tune' the brain. In women, the levels of estrogen and progesterone in the brain influence a number of skills, not just mental rotations but also many of the traits that males and females differ on, including memory, manual dexterity and verbal fluency. Some skills get worse in the presence of high levels of female hormones, but others improve. In men, testosterone seems to play a similar role, helping or hurting spatial ability, mathematical reasoning and perhaps other skills. The findings have implications not only for women on or off the Pill, but also for postmenopausal women receiving estrogen-replacement therapy, men getting testosterone treatments or taking steroids, and just about anybody else past the age of puberty.

This is a new side of sex hormones, one that is quite different from the role described over the past several chapters. In the womb and also after birth, sex hormones direct the development of body and brain, pointing them down either a masculine or a feminine path. Scientists refer to this role of sex hormones as *organizational*. When male hormones cause a penis to grow on a boy in the womb, that's an organizational effect. And when testosterone in a developing male rat's brain causes the SDN (sexually dimorphic nucleus, remember?) to grow five times larger than in a female, that's also an organizational effect. Organizational effects are, in general, permanent changes.

The other face of sex hormones, the one that Elizabeth Hampson saw in the women taking the Pill, is much more fleeting. In this *activational* guise, the hormones trigger certain temporary changes which come and go depending upon the hormone concentration. The most important and best studied example is the way that a woman's rising and falling hormones get her body ready each month for possible pregnancy – they signal the uterine lining to thicken in preparation for a fertilized egg and trigger other physiological changes in the body throughout the cycle.

But the hormones talk not just to the body – they act also on the brain. Testosterone levels affect a male's sexual desire, for instance, and researchers like Hampson are slowly accumulating a list of other ways that sex hormones affect behavior and thinking in humans. But the best place to see this activational role in action is still the rodent equivalent of the college student – the white rat. So let's start there.

A female rat ovulates every four or five days, and her sexual behavior is completely regulated by her hormone levels. She will be sexually receptive only at one particular point in her cycle – the few hours she is fertile – and will ignore males the rest of the time. Remove her ovaries, which produce the hormones that regulate her behavior, and not only will she stop ovulating but she will lose interest in males completely. That can be reversed with a shot of estradiol followed a couple of days later with a shot of progesterone. Within a few hours she'll behave just like a normal female rat on heat: her brain will have signalled ovulation to begin (although there are no ovaries to answer the signal) and she responds to a male who mounts her by arching her back, tilting her head and moving her tail out of the way. This hormone treatment has exactly the same effect on a male that was castrated at birth, making him a female for a day.

In causing this behavioral change, the hormones actually alter the microscopic structure of the brain, increasing the number of synapses (the communication points where one neuron passes a signal on to another) and making other changes in the parts of the brain that regulate sexual behavior. After estrus is past, the changes in the female rat's brain disappear, only to be created again a few days later during her next cycle.

Male rats don't have a reproductive cycle – they're ready for sex at any time – but their sexual behavior too is under the control of their hormones. Castrate a male to remove the source of his testosterone, and over a period of a month he loses his masculinity bit by bit – first he ceases ejaculating, later he stops thrusting, and finally he gives up completely, no longer even bothering to mount a ready and willing female. But put him on a two-week regimen of testosterone injections and he's soon back to his old self, regaining the pieces of his manhood (rathood?) in the reverse order that he lost them. Even females given enough testosterone will mount other females.

And although sex is the most obvious behavior under the influence of sex hormones, it's not the only one. Aggression is another well-known example. Although lab rats are not particularly aggressive animals, there are plenty of species in which testosterone does make males more aggressive. That's why farmers castrate some of their animals, for example – steers and geldings, with their testosterone-producing testes gone, are much gentler and easier to handle than bulls and stallions.

Several researchers have shown that female hormones increase nurturing behavior in lab animals. A female rat who has just given birth to a litter is immediately nurturing – licking them, carrying them back to the nest if they stray, and so on – and a female who has already had pups will also nurture pups that are not her own. On the other hand, virgin females and males will generally ignore rat pups at first (and males will often kill them), but both will gradually become nurturing with enough exposure to the pups. This change in behavior implies that rats can learn to be nurturing with practice, but hormones offer a shortcut: inject a virgin female with female hormones, and shortly she'll be treating the pups like a mother who has had many litters of her own.

Estrogen also seems to cause a female rat to lose her appetite, notes UCLA's Roger Gorski, one of the experts on sex differences in the rat brain. On the day that a female rat's estrogen levels are rising prior to ovulation, she'll move about much more than normal, he says, but 'she just nibbles at her food'. (The same thing works in rhesus monkeys – the higher the estrogen level, the lower the appetite.)

Given such details about rats and other animals, you might guess that sex hormones also have some influence on human sexual behavior – and you would be right. Men who don't produce enough testosterone, either because of castration or some medical disorder, often lose much of the normal desire for sex, although they remain capable of intercourse. (Castration entails removing the testes, but the penis remains intact and functional.) But put such men on a testosterone-replacement regimen, and they're back in business.

The picture is more complicated for women. As in men, higher testosterone levels in women are correlated with greater sexual desire and more sexual thoughts (women's bodies produce testosterone, too, just not as much as men's). But estrogen and progesterone seem to play a role as well. Some women, for instance, report a monthly increase in sexual desire around the time their estrogen levels are highest, just before ovulation. These women are a minority, however, and other women report peaks in sexual interest just before or just after menstruation. Researchers still haven't unraveled all of the factors at work here.

Of course, there is a big difference between rats and humans, which makes reasoning from rats a perilous practice. Put a male rat and a female rat in a cage and, if the hormones are right, they'll copulate immediately. There's no thinking about it – the hormones are in control. But humans' actions are influenced much more by learning and culture, and people have a conscious control over their actions that rats don't. Put a male and a female college student in a room, and even if the hormones are raging, they'll at least stop to close the door. 'We control our hormones,' Gorski says.

The role of socialization makes it tough to measure just how much clout hormones have in humans. If a mother is more nurturing to her newborn child than the father, is it her estrogen or just those early experiences playing with dolls, helping out with her little brother and babysitting the neighborhood kids? The jury is still out on this question. Recently, however, a preliminary verdict has been returned in a related case: aggression versus testosterone.

Because of the clear role testosterone plays in triggering aggressiveness in other mammals, researchers have long looked for a correlation in humans between testosterone and aggressive or violent behavior. On the surface it seems plausible. After all, most violent crimes in any society are done by males, who have testosterone levels several times higher than females, and the years when males are most likely to commit violent crimes – the mid to late teens and early twenties – are also the years when testosterone levels are at a peak. Furthermore, body builders and other athletes who take steroid hormones, which are similar to testosterone in structure and effect, often have personality changes, becoming more short-tempered and aggressive. In weight-lifting gyms, the effect is known as 'the 'roid rage'. (Besides being well-muscled, steroid users also tend to be super-sexed. One study found, for instance, that athletes who used anabolic steroids had more sexual thoughts, more morning erections, more sexual encounters and more orgasms than a control group of non-steroid-using athletes. Ironically, abuse of steroids often leads eventually to difficulty in achieving erections – which could create a scary situation for bystanders. Imagine a muscular steroid user, already prone to 'roid rages, who has been having an exaggerated number of sexual thoughts and who suddenly discovers he has trouble getting an erection. You probably don't want to be around.)

Despite the steroid evidence, however, the testosterone-aggression connection is not so simple as 'more testosterone = more aggression'. You might expect, for instance, that violent male criminals would have higher testosterone levels on average than the non-violent members of the male population, but the evidence is confused and contradictory. Some researchers have reported that violent male criminals do indeed have higher testosterone levels than non-violent offenders, while other studies have failed to find such a difference.

Part of the confusion probably stems from the way testosterone levels rise and fall in response to a situation. Fighting – and especially winning a fight – pumps up testosterone. This is true in male rhesus monkeys: when two groups of monkeys fight, testosterone levels go up in the winners and drop sharply in monkeys who are beaten up by other males. And it is true in humans. A study of members of the Harvard wrestling team found that testosterone levels increased during matches, but the winners had higher levels after the matches than the losers. Losing sent the hormone levels back down. Even something as simple as winning or losing a tennis match has been shown to lead to either a jump or drop in testosterone, depending on the outcome. So the testosterone-aggression connection is clearly not a one-way street.

Some researchers have looked for a correlation between testosterone levels and aggressiveness in normal, non-criminal subjects, usually by giving the subjects an aggression questionnaire that asked how they would respond in different situations. Several such studies have reported that, on average, the more testosterone a man has, the more aggressive he reports himself to be, but nobody really trusts such paper-and-pencil tests to give a good measure of aggression. After all, if you're a macho college guy, are you going to admit that if someone insulted your girlfriend you'd be more likely to turn and walk away than bop the guy in the nose? So it was a big deal when Mitch Berman and Stuart Taylor at Kent State University and Brian Gladue at North Dakota State came up with a way to compare testosterone and aggression more directly.

They used a rather complicated experimental set-up that combined elements from two standard types of psychology experiments – a competition between two people, and a shock-administration test in which the subject thinks he is setting the level of a shock being given to another person (although in reality there is no other person being shocked). And the researchers added a third, somewhat unusual feature to the mix – the subject would receive shocks himself during the course of the experiment.

It worked like this: thirty-eight men volunteered to take part in what they were told was a test of the effect of competition on reaction times and hormone production. One at a time, the subjects were sat down in a cubicle and told they would be competing with an (unseen) opponent in another cubicle to see who could press a button first when a light flashed. If the subject lost, he would receive a mild electric shock whose magnitude had been chosen by his opponent; if he won, his opponent would get the shock. Before each trial, the subject selected the level of shock to give to his opponent, choosing a number from 1 to 10. The shocks were not dangerous but they weren't fun, at least not at the upper end. Before the trials began, the experimenter had given a series of increasing shocks to the subject, who was asked to say when the shock reached a level that was 'definitely unpleasant'. This would be level 10, with level 9 being 95 percent of this maximum, level 8 being 90 percent, and so on. You wouldn't enjoy getting a lot of level 10 shocks.

The idea, of course, was to see how aggressive a subject would become in shocking his opponent. After each trial, a feedback light told the subject what level shock his opponent had chosen for that trial, so that the subject could use that information in deciding what level to pick for his opponent for the next trial. There were twenty-one trials, and before and after the trials the subject gave saliva samples to be tested for testosterone levels.

What the subject didn't know was that there was no opponent. He was playing against a computer, and the winner and loser of each trial had been set ahead of time by the experimenters so that each subject 'won' about half the time, no matter how fast or slow he was. Furthermore, the experimenters had also predetermined the shock levels so that the subject would think his opponent had started out choosing low shocks and gradually escalated until they were near the maximum at the end. By fixing things this way, Berman, Gladue and Taylor could guarantee that each subject was responding to exactly the same situation.

The goal was to look for a relation between the subjects' testosterone levels and the level of shocks they gave their imaginary opponents. Would high-testosterone males be meaner? This type of simulated electrical shock is the closest that psychologists can come to real aggression in lab experiments since they're unwilling to actually let a subject harm another person or animal, but it's a relatively good measure because the subjects truly believe they are hurting someone (or in the case of these rather mild electric shocks, at least annoying someone).

When the researchers compared their subjects' hormone levels with their shock settings, they saw that the high-testosterone males were significantly more aggressive. Splitting the subjects into two equal groups on the basis of their testosterone levels, they found that the group with the lower testosterone averaged setting the electric shocks at forty-two percent of maximum, while the higher group averaged fifty-seven percent. This could be important information if you're ever trading electric shocks with a man in the next booth: try to choose a partner with low testosterone, and avoid the guys with full beards and hair on their backs.

Unexpectedly, the researchers uncovered a second factor also involved in how aggressive the subjects were. Using a personality test, they determined whether each subject had a 'Type A' or 'Type B' personality. Type A people are the stereotypic overachievers – they're super-competitive in striving for goals, they try to control events, and they often get hostile when things don't go their way. Type Bs are more laid back, less controlling. Dividing the subjects into four groups based on whether they were Type A or Type B, high testosterone or low testosterone, the researchers found it was the Type A guys with high testosterone who were by far the most likely to set the shock at the maximum level. 'They like to hurt people,' Gladue says. By contrast, low-testosterone Type As were no more likely to choose the maximum shock setting than Type Bs. Perhaps, the researchers suggest, it is high testosterone in combination with a hostile, competitive personality that predisposes someone to react aggressively to anything seen as a challenge.

One way or the other, the experiment has clearly shown that men with more testosterone are likely to be more aggressive, at least when it comes to trading electric shocks and probably in other ways as well. Given what we know about testosterone and aggression in animals, it's not a surprising conclusion.

What is surprising is the growing amount of evidence that sex hormones manipulate more than just emotional behavior, such as aggression or sexual desire. They also shape those behaviors we think of as more human and less animal – the 'cognitive abilities', in the psychologists' jargon. Little more than a decade ago this would have been hard to believe, but now it's hard to deny, thanks in large parts to the efforts of Elizabeth Hampson.

Those efforts, which eventually led to her study on how the Pill affects spatial ability, began in the early eighties when she showed up at the University of Western Ontario for graduate studies in psychology. At the time, Hampson recalls, there had been 'tantalizing evidence' in the scientific journals that fluctuating hormone levels in women might affect their cognitive functioning in certain ways throughout the monthly cycle, but proof had been hard to come by. Furthermore, it wasn't something that most people wanted to believe. 'There was nothing conclusive,' she remembers, 'and a lot of people were saying that there were no cognitive fluctuations across the cycle.'

Her instincts told her otherwise, however, and she decided that part of the problem was that people just hadn't been looking at the right things. The logical skills to test, she reasoned, would be those that show sex differences. If varying estrogen levels did create a noticeable change in some mental ability, for instance, then that ability probably differed between the sexes as well since men's and women's brains are exposed to different amounts of estrogen. Conversely, if males and females didn't differ on a particular skill, it seemed unlikely that sex hormones played a big role in it.

Hampson's Ph.D. adviser was Doreen Kimura, the sex difference researcher who has traced out so many differences in the functioning of men's and women's brains. With Kimura's advice and guidance, Hampson set out to see for herself whether women really do experience monthly ups and downs in mental abilities. She decided early on that there would be two keys to the success of her research: using the right tests, and giving the tests to women at the crucial points in the cycle, when hormones were either at their lowest or at their highest.

Details of the monthly hormone fluctuation vary from woman to woman and even from month to month in the same woman, but the general pattern is the same. Estrogen levels are at a minimum during menstruation and through the first ten days or so of the cycle; then in a short preovulatory surge lasting just a few days they rise sharply to their highest point of the month and fall back down just as quickly; after ovulation, the estrogen starts a much more gradual climb that culminates in a second, smaller peak about a week before menstruation; and the estrogen then drops back down to a minimum as menstruation and the beginning of the next cycle approach. Progesterone's pattern is much like estrogen's, but without the peak just before ovulation – it is low during menstruation, stays low through ovulation, rises along with the estrogen to a high point a week before menstruation, and then slowly declines again.

One of the problems with testing women at different points in their cycles, Hampson says, is knowing exactly where in her cycle a woman is. The menstrual phase is easy to identify, of course, but pinpointing the preovulatory surge is much harder because it is so brief. So for her first study Hampson decided to compare women during menstruation and at the 'midluteal phase', the week or so toward the end of the cycle during which estrogen and progesterone both hit a peak. The midluteal hormone high lasts for several days, and although there's not as much estrogen then as during the preovulatory peak, Hampson could be relatively sure of testing women at the desired point in their cycle. She would merely ask the subjects to come in seven days before the expected onset of their next period, and then follow up with phone calls after the test to make sure the timing was right.

Just as important as the timing would be the types of tests. Since spatial ability shows larger sex differences than almost any other skill, Hampson decided to include a standard spatial test called the 'portable version of the rod-and-frame test'. A subject looks into one end of a rectangular tunnel about three feet long and a foot and a half wide. At the end of the tunnel is a rod, and the subject's task is to align the rod so that it is pointing straight up and down. (The subject doesn't turn it herself, but tells the experimenter how to adjust it.) The trick to this, Hampson says, is that the tunnel is tilted a bit so that its walls are slightly off from vertical, which makes it very hard to tell when the rod is straight up and down. Women tend to be off by an average of five or six degrees, two or three degrees worse than the average for men.

Hampson also decided to give the subjects three tests of manual dexterity, an area that women generally excel in. First they were asked to tap a telegraph key as fast as possible for several seconds, and the total number of taps was recorded electronically. Then they pushed pegs into a pegboard, with their score being the number of pegs inserted in a fixed time. Finally, they were trained on a 'manual sequence box', a clever device on which a subject must perform three actions in a row: push a button with the index finger, pull on a handle with all four fingers, and then twist the hand to tap the side of the thumb on a metal plate. They were timed on how long it took them to learn the sequence and on how quickly they performed it ten times in a row without a mistake. Women do both the pegboard and the manual sequence box faster than men.

Each subject was given the series of tests twice, once during menstruation, on the third to fifth day of her cycle, and once during the midluteal phase, seven days before the expected onset of menstruation. Half of the subjects took the tests first during menstruation and then again about six weeks later during the midluteal phase, while the other half went in the reverse order: midluteal, then menstrual.

After everyone had been tested, Hampson averaged the scores of her thirty-four subjects for both the menstrual phase and the midluteal phase, and compared them. This time it wasn't just 'tantalizing evidence' that she found. Her data offered solid proof that hormones were modifying performance across the menstrual cycle.

On the rod-and-frame test the women performed much better during the menstrual phase, when estrogen and progesterone were low. In the first testing session, the women who were menstruating outscored the women in the midluteal phase by an average of more than two degrees, or about three-quarters of the disparity that researchers see between males and females on this test, so the menstruating women scored almost as well as an average group of men would. The difference between menstruating and non-menstruating groups shrank when results from the second session were included, Hampson says, probably because of practice effects from the first session, but the scores from the menstruating group were still better. On the other hand, the women did better on the three tests of manual dexterity during their midluteal phase, when hormones were high. On the manual sequence box, the difference was about two-thirds the size of the usual difference between males and females.

Thus, Hampson and Kimura concluded, high levels of female hormones improve dexterity, a skill at which women excel, but inhibit spatial ability, at which males excel. And they didn't stop there – they went on to draw the logical conclusion about sex differences. In general, women have much more estrogen and progesterone in their bodies than men, so perhaps the differing levels of sex hormones in males and females are responsible for some of the differences in abilities between the sexes, the two researchers suggested. Estrogen may give women an edge over men in manual dexterity while it puts them at a disadvantage in spatial reasoning.

Hampson realized that such provocative claims would be closely scrutinized, both by fellow sex difference researchers and by others with a stake in the results, such as feminist scientists. 'To make these sorts of claims you have to have fairly airtight evidence,' she says. So even before the first study was published she was back at work on a larger one to check the original results.

This time she recruited a slightly larger sample – forty-five women instead of thirty-four – and added more tests to her battery. The results were the same. During the midluteal phase, when female hormones were high, the women did better on tests that females normally excel on: manual co-ordination, articulation (speaking rapidly without mistakes), and verbal fluency (quickly coming up with appropriate words, phrases or sentences). During the menstrual phase, when hormones were low, the women did better on the spatial reasoning tests: the rod-and-frame test again, an embedded figures test, and a type of mental rotations test.

This was encouraging, but Hampson wanted even more confirmation, so she designed a study to compare women during menstruation and the preovulatory peak. This would be harder to carry out, but it would have the added advantage that during the preovulatory surge only estrogen is up – not progesterone – so if she found an effect she could say that it was due to estrogen alone. To make sure that the women really were tested during the brief preovulatory peak, she gave each of them blood tests and kept results from only those subjects who showed high estrogen levels. Of eighty-eight women who signed up for the experiment, Hampson had fifty in her final group. The results were nearly identical to those in the second study. It was estrogen, not progesterone, that was producing the effects.

More recently, Irwin Silverman of York University in Toronto and Steve Gaulin at the University of Pittsburgh have done similar studies on menstruating and non-menstruating women and also included males in the same experiment for the sake of comparison. They focused simply on the Vandenberg mental rotations test because it gives such large and consistent sex differences, and when they analyzed their data they found that – as Hampson and Kimura had suggested – high estrogen levels in females do seem to account for a large part of the difference between men and women, at least on this test.

In one group of eighty-six males and one hundred and eight females, the average male score was 14.9 correct, and the average score for a non-menstruating female was 7.5, or about half the male score. But among the women who had been menstruating when they took the test and thus were at the low point of the estrogen and progesterone cycles, the average score was 10.8. The sex difference between males and menstruating women was thus little more than half the sex difference between males and non-menstruating women.

Silverman and Gaulin's results are very similar to the study that Hampson did with Szekely on women taking birth control pills. Among the contraceptive users that they tested, Hampson says, those who were in their menstrual week scored twenty-five to thirty percent higher than those taking the Pill the day of the test. The synthetic estrogen contained in the Pill appears to have the same effect on spatial ability as natural estrogen, diminishing scores on the Vandenberg test by a noticeable amount.

Spatial ability, manual dexterity, articulation and verbal fluency – Hampson admits that many people, particularly women, may not be happy to hear that science has found that all of these abilities vary in women according to the point in their menstrual cycle. Indeed, it seems uncomfortably close to what some have claimed about women in the past – that their monthly cycles make them erratic and unpredictable. Isn't this just a new way of saying that a woman is a slave to her hormones?

Not at all, Hampson reassures. The practical implications for an individual woman are probably quite small. Yes, tests that focus exclusively on spatial ability or verbal fluency may detect the influence of hormones, but few things we do in real life depend on only one skill. Most of the challenges that people face on a daily basis – solving practical problems, interacting effectively with other people, making decisions based on incomplete information – involve integrating many different mental functions, only a few of which may be influenced, for better or worse, by hormone levels. So for most tasks, a woman's estrogen level will probably have less effect than what she ate for lunch. In general, the influence that hormones wield varies considerably from woman to woman and probably from month to month, and it can usually only be measured by the types of specialized tests that Hampson and other sex difference researchers use.

Indeed, it may only be certain women who are affected by estrogen variations across their cycle, according to work by other Canadian researchers. Barbara Sherwin at McGill University in Montreal and her student Susana Phillips studied how short-term memory varied in twenty-five women between the menstrual phase and luteal phase (days 19 to 24) of their cycles. On one test, for example, they showed the subjects several simple drawings and later asked the subjects to reproduce them from memory. Sherwin found that as a group the women did better on this task in the luteal phase, when estrogen and progesterone levels were higher. But when she looked more closely she saw that it was only some of the subjects. Twelve of the twenty-five did better in the luteal phase, ten did about the same at both times, and three actually scored better during the menstrual phase.

When Sherwin focused on just those twelve women whose performances improved with the extra estrogen and progesterone of the luteal phase, she found a pattern between hormone levels and performance: the more progesterone during the luteal phase, the better a woman did on the visual recall test. But it was only true for those twelve, and some of the other thirteen women produced just as much progesterone as these dozen without any improvement in their performance. It may be, Sherwin suggests, that some women are more sensitive or responsive to changes in hormone levels than others, and these are the women whose performance changes on various psychological tests across their cycles.

This is consistent with what is known about mood changes during the cycle, Sherwin notes, since only about twenty to forty percent of women report physical or emotional symptoms when estrogen and progesterone drop during the premenstrual phase. This premenstrual syndrome, or PMS, is one of the most contentious areas of the research into women's hormone fluctuations, partly because it has been the subject of so much hype. The effects of PMS have been exaggerated and oversold by authors trying to sell books, by drug companies trying to sell cures and even by defense attorneys trying to sell excuses to a jury. (It worked at least once – in England, when a woman who had killed her boyfriend by running over him with a car was given a conditional discharge from jail after blaming it all on PMS.)

Despite the hype, there is no consensus on exactly what PMS is or what causes it, although sex difference researchers have often noted that there does seem to be a connection in some women between female hormone levels and mood – in general, high levels bring better moods and puny amounts can leave you feeling low. When Hampson did her study of women during menstruation and the preovulatory phase, for example, she gave all the subjects a questionnaire about their mood before testing their spatial ability and manual dexterity. On average, Hampson found that the fifty women in the study reported being more tense, depressed, angry, tired and confused as well as less vigorous during the menstrual phase – or, if you prefer, less tense, depressed, angry, tired and confused and more vigorous during the preovulatory phase. The surge in estrogen just before ovulation does appear to give some women an emotional and physical boost, Hampson concludes.

Whatever PMS is or does, Hampson, Sherwin and other researchers don't believe it has anything to do with the mental changes they see in women across their cycles. Yes, some abilities, such as short-term memory, manual dexterity and verbal fluency, do seem to peak at the same time of the month that women report their best moods – that is, when estrogen and progesterone are high. But the researchers normally give their subjects mood questionnaires before administering the tests and then check to see if the women who score better on the tests are those who report feeling better. Consistently, this is not the case.

And of course, if the differences in test performance are due to mood swings, it's hard to explain why women score better on spatial tests when their estrogen levels – and their vigor and sense of well-being – are down. Hampson says she certainly can't come up with a plausible explanation.

Although much of the research into the effects of female hormones on test performance have been done on normally cycling women or women taking birth control pills, Barbara Sherwin and her student Susana Phillips have looked into women receiving estrogen-replacement therapy. For one study, Sherwin and Phillips recruited nineteen women, average age forty-eight, who were going to have their ovaries surgically removed. They randomly divided the women into two groups – ten who would be given estrogen replacement after surgery plus nine who would receive a placebo. None of the subjects knew who was getting estrogen and who wasn't, so their expectations could not influence the experiment. Sherwin and Phillips gave the women a series of tests shortly before their surgery and again two months afterwards.

After scoring the tests, they found that although there were no differences between the groups before surgery, the women who received estrogen replacement after surgery performed better on several tests of memory than the ones who got only a placebo. On one of the tests, for instance, the subject first heard a list of pairs of words ('necktie/cracker', 'fruit/apple' and so forth) and then were asked to remember the second of a pair when given the first ('What word went with necktie?'). After surgery, when estrogen levels had dropped sharply, the placebo group did significantly worse, but the women who received estrogen replacement did just as well as they had before surgery. And on a paragraph-recall test, where the subjects had to repeat from memory a paragraph that had been read to them, the women who had estrogen replacement actually scored better after surgery than before, while the placebo group's performance did not improve. The improvement over pre-surgery scores could have been due, Sherwin says, to the fact that estrogen levels in the women who had hormone replacement were actually about five times higher than they had been before surgery.

This experiment may have implications for postmenopausal women, whose ovaries stop producing estrogen, Sherwin says. Doctors hear many complaints of memory problems from these older women, and part of the explanation, she says, may lie in the much lower levels of estrogen these women's bodies are getting. But her study implies that it is only verbal memory that suffers in women with low estrogen; she saw no effect on visual memory – remembering pictures and drawings.

Sherwin echoes Hampson in emphasizing that these hormonal effects on memory – while scientifically interesting and important – probably have few practical implications. There may be some postmenopausal women whose memories could be helped with hormone replacement therapy, she says, but the mild declines she saw in test scores of women after surgery 'were not, to the best of our knowledge, associated with any impairment in daily activities.'

The real importance of these results to her, Hampson says, is that they imply that sex hormones can actually communicate with the cortex – the thinking part of the brain. It's not much of a surprise that the emotional areas of the brain, those that control sexuality, aggressiveness and other passions, can hear the call of these hormones, but a decade ago few would have thought that sex hormones could make themselves heard in the parts of the brain with a cooler head, so to speak. On the other hand, knowing what we do now about the influence that sex hormones wield in the developing brain, the effects later in life probably should have been expected. Neuroscientists have discovered that many of the same chemicals that guide the fetal brain's development play important roles in the day-to-day workings of the adult brain, so given the work of people like Sheri Berenbaum, whose studies on CAH girls have proved that sex hormones in the womb influence the thinking parts of the brain, Hampson's findings fit right in. In that sense, the influence of sex hormones on mental rotations or remembering words is just one more part of our hormone heritage.

* * *

What about men? Surely they're not immune to this heritage. After all, if testosterone in the womb molds such mental skills as spatial ability, shouldn't it play a role later in life, just as estrogen does?

The answer is that testosterone does seem to play a role in men somewhat comparable to estrogen's role in women, but the details are hazier because much less research has been done. This isn't a case of discrimination, however – it's simply that the effects of hormone fluctuations are easier to study in women. Testosterone does go up and down throughout the day in men and also varies from season to season, but the changes are rather gentle compared to women, whose estrogen levels may be five or six times higher just before ovulation than during menstruation. Some men have higher-than-average testosterone levels, others lower, but past puberty the levels stay relatively stable in each individual male.

Despite this, a few researchers have tried to track cognitive abilities in men as hormones go up and down, and the preliminary evidence is that some skills – spatial abilities in particular – may be just as sensitive to fluctuations in testosterone as they are to fluctuations in estrogen. In one experiment, Doreen Kimura gave mental rotations tests to men at two points during the year – in the spring, when testosterone levels are low, and in the fall, when they are high. The males tested in the fall did much worse than those in the spring, Kimura found, indicating that high levels of testosterone may hinder spatial ability much as high levels of estrogen do. The effect is large enough that men tested in the fall, when testosterone is high, seem to do no better on the test of mental rotations than women who take it during their menstrual periods, when estrogen is low.

If you've been paying attention, the connection that Kimura found between testosterone and spatial ability probably seems backward. Since men outscore women on spatial ability, you might guess that the more testosterone a man has, the better he would score on mental rotations and other spatial skills. But other experiments comparing the levels of male hormones with test performance indicate that Kimura may be right – too much testosterone may actually hamper some mental abilities. It's one of the real ironies of sex difference research that the more 'masculine' males – in the sense that they have more testosterone – may actually have less spatial ability than their low-testosterone peers.

Most studies of testosterone's influence take a different approach from Kimura's. Instead of measuring how men's test scores vary across time as their testosterone levels fluctuate, the studies simply compare different men. Since some men have two or three times the level of male hormones in their systems as others, it's possible to test a large number of men, measure their hormone levels, and then see if there is any correlation between testosterone and performance. This approach has the advantage that you can include women in the sample and see how their performance and testosterone levels relate, too. The results have proved surprising.

In 1990 Catherine Gouchie, another student of Kimura's at the University of Western Ontario, recruited forty-two men and forty-six women to help her follow testosterone's trail. She gave each of them a battery of tests and also took saliva samples which would later be analyzed to determine each person's testosterone level. The collection of tests included several on which males usually excel (mental rotations, a second spatial test called paper folding, plus a test of mathematical reasoning), some on which females excel (perceptual speed tasks, such as quickly determining whether two pictures are identical, and some basic math problems), and a test with no sex difference (vocabulary).

To analyze the relationship between male hormones and test scores, Gouchie and Kimura split the men and women each into two equal groups based on their measured testosterone levels. That produced four groups: high-T men, low-T men, high-T women, and low-T women. The average testosterone concentrations for the groups, measured in picograms per milliliter, were 212, 138, 69 and 33, so even the high-T women averaged only half the testosterone of the low-T men. Then Gouchie and Kimura analyzed the average test score for each group.

Among the men, the results agreed with Kimura's study on spatial ability across the seasons. The men with low testosterone levels scored higher than the high-T group on every test that normally favors males. The difference was particularly large on the paper-folding test, but it was also evident for mental rotations. In addition, low-T men outperformed their high-T peers on the test of mathematical reasoning.

But the women's scores showed just the opposite pattern: the high-T women significantly outscored the low-T group on the paper-folding test, and they had a smaller advantage on mental rotations and mathematical reasoning.

At first glance, this may not seem to make much sense – if less testosterone is good for men on these tests, why shouldn't it be good for women too? But Gouchie and Kimura have faith in their results, especially since they conform with an earlier study done by Anne Petersen. Petersen studied teenaged boys and girls, comparing their body types with their scores on spatial and verbal tests. She found that the more masculine-looking boys – the ones with more facial hair, broader shoulders and better muscle development – were better at verbal than spatial tasks, and the less masculine ones excelled at spatial over verbal. Among the girls, the ones with large breasts and hips and less muscle development were not as good at spatial tasks as the ones who were more androgynous in appearance.

Since body types are a rough indication of hormone levels, Petersen's results seem to point to the same connection that Gouchie and Kimura found: high-T males have poorer spatial skills than low-T males, and the opposite is true for females. Of course, this is true only on average and not necessarily for any given individual – very masculine-looking men can have high spatial ability, as can very feminine-looking women.

One more bit of evidence comes from a series of recent studies on musicians and testosterone level by a German researcher, Marianne Hassler. In one study, for instance, she found that male composers had testosterone levels at the low end of the scale for men, while female composers were at the high end for women. Hassler concludes from her experiments that there seems to be an optimal range of testosterone levels, and that either too much or too little testosterone is not conducive to creative musical talent.

Gouchie and Kimura think the same thing is going on with spatial ability – there appears to be an optimal level of testosterone for spatial ability and perhaps also mathematical reasoning. The high-T men in their experiment had too much testosterone, while the low-T women and even the high-T women had too little. The optimal level appears to be approximately what was found in the low-T men, since they had the highest scores of the four groups on both of the spatial tests as well as on the mathematical reasoning test.

On the other tests, Gouchie and Kimura found no significant relationship between testosterone levels and performance, either in men or women. It appears that male hormones have little or no influence on skills that women do just as well as or better than men.

All this implies that although both men's and women's brains are influenced by the levels of sex hormones in their bodies, the effects are different for male and female hormones. In women, fluctuations in estrogen and progesterone seem to affect many of the skills which exhibit sex differences – 'feminine' abilities such as verbal memory, manual dexterity and verbal fluency are best when female hormones are high, worst when they're low, and the 'masculine' trait of spatial ability is the reverse. Testosterone levels, however, seem to be correlated only with those skills in which males hold an edge, which indicates that testosterone may work on the brain in a different way than the female hormones do.

Researchers are still debating and speculating over what that difference may be. Gouchie and Kimura suggest, for example, that part of the answer could be that estrogen and testosterone fine tune different parts of the brain. Estrogen may enhance performance of the left side of the brain, which controls many speech functions, while testosterone may improve the functioning of the right hemisphere, which plays a dominant role in spatial ability.

But the difference may be more fundamental, reflecting the different roles that estrogen and testosterone play in creating man and woman. In some sense, estrogen is the more essential hormone, necessary for basic functioning of the brain, while testosterone's role is supplemental, adding features that the brain can do without.

Estrogen's influence on such things as verbal memory may arise from its ability to create nerve connections in adult brains, Barbara Sherwin says. A series of experiments on rats done in Bruce McEwen's lab at Rockefeller University in New York City has shown that estrogen increases the density of synapses and the number of dendritic spines on neurons, both of which are points at which one neuron communicates with another. Thus, Sherwin suggests, when estrogen-replacement therapy improves memory in postmenopausal women, it may be thanks to its effects on neurons in the hippocampus, the part of the brain where memories are formed.

A similar action might explain why higher estrogen levels lead to higher scores on articulation, verbal fluency and manual dexterity, such as Hampson found in her studies of women across their cycles. It's harder to understand why high estrogen levels should hamper women's performance on spatial tasks, Hampson admits, but it's conceivable that estrogen has different effects on different parts of the brain, sometimes helping, sometimes hurting.

The connection between testosterone level and spatial ability might also arise out of some action of the hormone on brain cells, but it may not be that simple, to judge from the study of men with idiopathic hypogonadotropic hypogonadism described in Chapter 4. These men have abnormally low levels of male hormones from before birth and consequently very low spatial ability, but when they are given hormone-replacement therapy to bring their testosterone levels up to normal, it doesn't improve their spatial skills. Thus something more is needed than just the right level of testosterone to have good spatial skills.

That something is likely to be the proper hormone environment during development, Gouchie and Kimura say. There is probably an optimal testosterone level for creating spatial ability in a brain as it is forming – too much testosterone or too little, and spatial ability won't be as good as it could be when that person becomes an adult. If so, much of the testosterone–spatial ability connection in adults probably traces back to before they were born, the researchers say. Since a person with high (or low) testosterone levels in the womb will probably have high (or low) testosterone levels as an adult, the advantage that low-T men have on spatial tests may be little more than an echo of the hormones they were exposed to in the womb.

CHAPTER 9

Nature/Nurture

For eight chapters and some two hundred pages we've heard about hormones and how they push males and females down different paths, but no one – certainly none of the dozens of scientists interviewed for this book – claims that hormones tell the whole story. Boys and girls, men and women live in a world in which the sexes are treated very differently from birth. The first question one asks of new parents is, 'Is it a boy or a girl?' and it snowballs from there. Boys are given blue sailor suits and blocks to build with, while girls get flowery pink dresses and dolls that they learn to clothe and feed and nurture. In school, boys generally receive more attention from teachers, while girls are expected to behave better. As adults, nearly every aspect of life is colored by one's sex.

Over the past several decades, researchers have been exploring how males and females are treated by society, and they have accumulated a wealth of scientific evidence that the two sexes are indeed faced with environments that are different in many ways. For instance, study after study has found that, even when teachers insist they treat their male and female students just the same, the boys in a classroom get called on more often and praised more lavishly for correct responses. Teachers also send out hidden signals in their choices of students for various tasks, such as being more likely to pick boys to help with such chores as cleaning blackboards or dusting erasers.

Society as a whole has certain fixed ideas – or stereotypes – about what to expect from males and females. In general, people expect males to have more of the traits that psychiatrists call 'instrumental': competence, task orientation, dominance and so on. Females are expected to have 'warmer' traits such as compassion, expressiveness and nurturance. It's not exactly illegal for a man to be compassionate or a woman to be domineering, but many people see such behavior as somehow not quite right. The effects of such expectations are particularly severe in adolescence, when boys and girls are so sensitive to the opinions of their peers, but they don't disappear in adulthood.

You don't have to take my word for it, though. It's easy to see it for yourself. Go to a movie, look at the advertisements in a magazine, or just watch people in an office or on the street, you can't miss it – different things are expected from men and women, and people respond to the sexes in different ways. And of course, it's human nature to fall into the patterns that people expect of you. Few have the desire or the strength to constantly resist the expectations of their peers.

But if the environments that boys and girls grow up in – and that men and women live in – are so dramatically different, could this not account for the dissimilarities between the sexes? Do we really need to bring hormones into the picture? Twenty years ago perhaps a majority of psychologists would have said the environment is enough. Even today many believe that most of the important differences between men and women would disappear if only society would treat males and females the same. This, roughly speaking, is the nurture side of the so-called nature/nurture debate, a debate that has been raging not only around sex differences but around the entire question of 'What is human?' for several centuries.

I recall vividly one particular episode in this ongoing debate. It was the first evening of a NATO-sponsored conference on sex differences held at a hilltop château in the south of France. The weather was perfect and the scenery magnificent, with the château overlooking tree-lined fields of wheat and sunflowers that extended to the horizon. We had finished drinks and hors d'oeuvres and proceeded to dinner, and at the table I found myself seated between Camilla Benbow of Iowa State University and Rhoda Unger of Montclair State University in New Jersey. They had only just met but, perhaps because of the wine and the atmosphere, they were soon arguing like old friends.

Benbow described her research on precocious kids, the seventh-graders who outperform high school seniors on the SAT. On the mathematics section the boys consistently score better than the girls, and the disparity becomes larger as you look at kids with higher and higher scores. Since the boys had taken no more math courses than the girls, because the SAT included problems that neither sex had been taught how to solve, and since the sex difference had remained steady for fifteen years, Benbow concluded that most or all of the male advantage was innate, probably caused by the brain being exposed to testosterone during development.

Now wait a minute, Unger said. There are plenty of other things that might cause it. Boys are encouraged more than girls to excel in math, and maybe the boys are working math problems outside of class and getting practice, while the girls are discouraged from spending too much extra time on math. Girls hear from an early age that boys are better at math, so you may be seeing nothing more than a lack of self-confidence. Or perhaps it's simply that the girls don't try as hard on the tests because they've learned it's better to let the boys win.

No, Benbow said, I've investigated encouragement from parents and I don't see a difference between the boys and girls. These are the top students from each school, they're good and they know it – there's no lack of confidence. And if girls are letting the boys win, how do you explain that girls generally get better grades in math? Not only that, but the best math student in any given school will often be a girl. The pattern emerges only when you look at dozens or hundreds of schools.

Furthermore, she said, shifting ground slightly, studies of identical twins who have been raised apart find that their IQs are about as similar as identical twins raised together. This implies that environment has very little to do with a person's intelligence, so it's pretty much a waste of time coming up with all sorts of explanations for how environment might explain the sex difference among the mathematically precocious youth.

Not so, said Unger. There's more than one way to interpret those twin studies. Other studies have shown that the way people are treated depends very much on their appearance, and since identical twins look alike, they are probably exposed to the same type of environment wherever they are. Boys and girls don't look alike.

And on it went. The thing that struck me most about the debate was how it was possible for two rational, well-thought-out scientists to look at the same phenomenon and blame it on such diametrically opposed causes, each throwing out studies and statistics to support her own view. Although the nature/nurture debate is often exaggerated or oversimplified almost to the point of caricature, particularly by reporters trying to make it comprehensible to the general public, make no mistake: there is a real fight between those who think biology plays a significant role in the creation of sex differences and those who think it doesn't.

Part of the fight is simply a clash of scientific cultures. Scientists are smart, well-educated people, but they're human – and humans inevitably have biases and prejudices, some well-founded, others not. The biggest bias may lie not in a scientist's politics but in his or her choice of 'tools', the theories and investigative methods used to understand the world, and in the fact that most scientists come at a problem from one particular direction. Social psychologists, for instance, are trained to look for how the environment leaves its mark on a person. Their schooling, their interaction with peers, the sorts of experiments they do, the types of journals they read – all reinforce the idea that the world should be approached from the environmental perspective. If the only tool you have is a hammer, everything looks like a nail.

And once you've invested years of your life looking at things from one direction, you can get set in your ways. Put yourself, for instance, in the place of a (hypothetical) researcher who has written half a dozen papers explaining how sex differences in toy preferences are due to subtle parental pressures and who has just read Sheri Berenbaum's report that CAH girls – normal-looking girls exposed to testosterone in the womb – prefer boys' toys. Do you say, 'Oh, well, I guess it's all caused in the womb, so I can throw five years of my work in the wastebasket'? Not likely. If you don't ignore Berenbaum's paper altogether, you're going to look for any flaws you can find with it, anything to indicate that your own work is still valid.

But I don't mean to pick on social psychologists – limited perspective is a problem for every field. Although many scientists do try to stay abreast of developments in other fields, it's hard. There is so much going on in twentieth-century science that it's a full-time job to keep up with just a small piece of it, and researchers tend to find themselves focusing most or all of their energy on their particular areas of specialization.

This type of cultural divide, with scientists from different fields talking past each other, breeds debate in many areas of science, but the nature/nurture clash is particularly fierce thanks to the presence of a second factor: political ideology. Sometimes the influence is quite direct. When Anne Fausto-Sterling wrote her influential *Myths of Gender*, she was very open about her ideological prejudices. For political reasons, she wrote, she believes in equality between the sexes (meaning 'sameness') and therefore she demands an extremely high standard of proof for any claims that biology contributes to sex differences.

But more often it's indirect. At some point in his or her career, a scientist chooses which tools he or she will use to understand the world, and it makes sense to choose those tools you believe – consciously or subconsciously – will be most useful. If you think biology plays little or no role in how humans behave, you're unlikely to go into a field that studies the effects of hormones in the womb on behavior. You'll go into social psychology or a related field. If you believe biology plays a big role, you'll pursue a different career path, biological psychology or neuroendocrinology or behavioral genetics. And this self-selection process is self-reinforcing: once you start work in an area, you hang around with other people who see the world the same way you do, strengthening your initial impression that this is the right approach.

And mixed in with one's gut feelings about nature and nurture are the political implications of those beliefs. For three hundred years the idea that nurture is the determining factor in human actions has been used to argue for a variety of social programs – if we can only change people's environments, the argument goes, we can make a better humanity: reduce crime, fight poverty, get rid of the sex difference in math. This creed is usually associated with the politics of the left. On the other hand, if nature is responsible for much of what we are – heavily influencing such things as intelligence, personality traits, perhaps even propensity to commit crimes – then there are limits to what social programs can do to reduce the individual differences between people. This conviction is often thought to fit better with the politics of the right. (It's important to note, however, that although these are the traditional attitudes, a belief in nature does not necessarily imply conservative convictions, nor a belief in nurture liberal ones. This is clear, for instance, in the current debate over homosexuality. Apparently because they believe that finding a genetic basis for homosexuality would boost the cause of gay rights, political conservatives are unhappy with attributing sexual orientation to nature, while liberals are embracing the idea of biological determination – in this one area, at least. Furthermore, many of the researchers investigating the biological roots of sex differences are themselves politically liberal, and they certainly don't believe that their work supports conservative positions.)

In short, the nature/nurture debate is much more than simply a scientific argument. It is a clash of cultures, of ideologies, and of politics.

None the less, despite all that, it is still possible to cut out the extraneous stuff and just look at the science. It is possible because, in spite of all the pressures, most researchers still do careful, objective experiments that can be reproduced by other scientists, and because the price of admission into the scientific lore remains replication – if an experiment doesn't give the same answers the second time, and the third and the fourth, it's thrown out on its ear. Ironically, much of the science in this area is probably more trustworthy than results in other fields because it is battle-hardened. Researchers who claim to find biological influences in sex differences know they'll be challenged strongly by others who distrust such results. 'To make those sorts of claims you have to have fairly airtight evidence,' says Elizabeth Hampson, the researcher at the University of Western Ontario who found women's spatial skills and manual dexterity rising and falling with their hormone levels.

But the most important reason that the science of sex differences rises above politics and ideology is that very often (though not always) the scientists themselves do. Many of the researchers who are exploring the echoes of the womb are children of the sixties and early seventies – politically liberal citizens who came of age at a socially ambitious time when it is was widely believed the environment plays the major role in fixing a person's course. It would certainly be more comfortable and more convenient for them if the evidence said that all sex differences are created by the environment. But it does not, and as scientists they follow the facts.

Some critics of this research – including some scientists – have argued that you can't trust the science done on such a controversial topic because the work will inevitably reflect and be colored by the researchers' personal biases. To the contrary, research on the biological underpinnings of sex differences is probably stronger and more trustworthy because of the stakes, with the researchers' own natural skepticism of their results providing a balance to the science that would be hard to get any other way. And thanks to these scientists, we're starting to get some answers to the nature/nurture conundrum.

The first thing you learn when you wade into the sex difference nature/nurture debate is that it's useless to try to determine a winner. Each side can claim its victories. The biology bunch may have CAH girls, but the socialization camp can point to the experience of science itself. The increasing number of women earning Ph.D.s over the past twenty years proves that the traditional male dominance in that field was due to socialization, not any failing in women.

But the second – and more important – thing you learn is that it's silly to even think of the debate as a contest, chalking up points to nature for this and to nurture for that. They're not competitors, they're collaborators. The two are inextricably linked, so closely associated that it is truly impossible to tell where one starts and the other leaves off. Nature creates the brain, setting limits on what and how a brain can learn. Nurture comprises the learning itself, without which the brain is nothing, and the effect of nurture is to cause changes in the brain itself.

To take an extreme example, if a young child develops a cataract that prevents him from seeing through one eye, he risks becoming permanently blind in that eye unless the cataract is removed quickly. The reason is that the nerve cells in the baby's visual cortex are still growing and making connections with other nerve cells, and without stimulation by light from the outside, the neurons that would normally process signals from the one eye rewire themselves so that they work for the other eye instead. After a certain critical period, the wiring is pretty much fixed, so that the eye with the cataract will never see. The lesson here is that the development of the visual cortex depends not just on some pre-programmed biological blueprint of the brain, but also on input from the environment – nature and nurture working together.

This interdependence between nature and nurture is no secret among scientists. Even the most ardent biology proponent recognizes that a human is at least in part a product of his or her environment, and the most hard-headed socialization advocate admits that the brain is something more than a blank slate waiting to be written upon. But it's usually easier and more convenient for a scientist to try to isolate one factor or the other, so it's a rare researcher who actually looks at biology and environment at the same time. Janice Juraska is one such rare researcher.

'I'm fond of showing that things are complicated. I have a hard time latching on to simple explanations.' Juraska is speaking from a sofa in a room near her office in the psychology department of the University of Illinois in Champaign. It is Saturday. I arrived around ten in the morning at her office only to find a note on the door directing me to this room, which she uses for analyzing data and meeting with students. She cannot hold the interview in her office, she explains, for she has hurt her back and cannot sit or stand for too long, but she is happy to talk with me here where she can lie down. 'Do you mind?' she asks. 'Not at all,' I reply. I do discover, however, that sitting in a chair with my notepad and asking questions of her on the couch gives me an almost overwhelming urge to say, 'Now, tell me again why you hated your father.'

Juraska, another of the young women who have redefined sex difference research over the past decade, has revolutionized thinking on how the environment affects male and female brains, and although her work is done in rats, the human brain is never far from her thoughts. 'Sex differences are interesting because of humans,' she says. Her work reveals a complex interplay between nature and nurture and – assuming that it can be applied to humans – implies that even if boys and girls could be raised in identical environments they would never turn out identically. Further, it may be the case that the richer and more stimulating an environment is, the larger the sex differences become.

These are uneasy ideas for many people, Juraska says. 'When I'm talking to people at a party or something, I can always tell when somebody is just a little uncomfortable or appalled with what I do. They'll ask, "Aren't you afraid it will be misused?"' Yes, she says, she does worry about that sometimes and she keeps a relatively low profile because 'I would rather be underused correctly than misused.' But, she says, it's impossible to go back to the days of believing that the sexes could be identical if only the environment were right. 'We know too much to turn back.'

Juraska first began studying the effects of the environment on rats as a graduate student in the early 1970s. At that time researchers had known for a quarter of a century that rats raised in stimulating environments do much better on mazes and other tasks than rats raised in isolation. Something about growing up in a large area with other rats and plenty of toys to play with gives rats an advantage over those who are kept alone in small wire cages, and in the 1960s and 1970s researchers found physical differences in the cortex, the thinking part of the brain, that seemed to correspond to these behavioral differences – the rats raised in the complex environment not only had bigger neurons, but the synaptic contacts between the neurons were larger and the neurons had richer branching structures. (A neuron is structured much like a tree, with a bunch of branches, or dendrites, that collect signals from other neurons and a long root, or axon, that is used to pass messages out.) Here was clear, direct evidence of how rearing affected biology: an interesting environment stimulated the brain to grow larger neurons with more connections.

But, Juraska says, during more than twenty years of studying the effects of environment on the brain, researchers had focused on only male rats. 'I was always rather taken aback that they were just running males and no one was concerned about whether the same was true for female brains.' The standard explanation, she remembers, was that female brains – because of their hormone fluctuations – were too messy to deal with and, anyway, female brains would be no different from male brains. Juraska had her doubts, however, and after getting her doctorate in 1978 from the University of Colorado, she began to study the effects of isolated versus enriched environments on both male and female brains.

What she found surprised everyone. The female brains were not simply copies of the male brain. They had their own pattern of response to the environment. In her first series of experiments Juraska measured the 'total dendritic length' – the sum of the lengths of all the branches of a neuron – for three types of neurons in the rat cortex. In male rats, as expected, the numbers were consistently larger for rats raised in interesting environments. The neurons were growing either more or longer branches in response to more input from the outside world. But in two of the three types of neurons in female brains, Juraska found no difference between the isolated group and the enriched group. More stimulation was not creating more neuron growth in the females.

Clearly the environment was affecting male and female brains differently, so Juraska took the obvious next step of comparing the sexes directly to see exactly what the disparities were. In the brains of rats raised in isolation she saw no difference between the males and females, but in those from the enriched environment, the neuron 'trees' were larger in male brains than female brains. *Raising the rats in a rich environment had created sex differences that did not exist without the stimulation of this setting.* Nature had designed a sex difference that only developed with some help from nurture.

After that initial, surprising finding, Juraska looked to other parts of the rat's brain to see if similar things happened in other places. They do. In both the corpus callosum and the hippocampus, male and female rats' brains develop differently depending on whether the rats are raised in isolation or an enriched setting. Juraska's papers are some of the most difficult to follow in all of sex difference research because they deal with the fine structure of individual neurons, a topic foreign to all but brain specialists, yet for the same reason her results are some of the most compelling in the field. The secrets of how the brain works are to be found at the level of the neuron, and when sex differences appear here you can be sure it's not a coincidence or a meaningless accident.

In the corpus callosum, the band of tissue that connects the two halves of the brain, Juraska uncovered a situation that was the reverse of her original study of the cortex. In the rear third of the corpus callosum, there was no sex difference – it was larger in the rats from enriched environments for both males and females. But in the front and middle thirds, a stimulating environment increased the callosum's size for females but not for males. And when Juraska examined the individual axons in the corpus callosum, breaking them down into myelinated and unmyelinated (myelin is a material that covers some axons to help electrical signals travel faster along them), she discovered a complicated pattern of differences in how the environments affected the brains of the sexes. For instance, being raised in an enriched environment expanded the size of the myelinated axons in males but increased the number of those axons in females.

The different patterns of response to the environment probably have their roots in the different hormones that male and female brains are exposed to during development, Juraska says. In the hippocampus, for instance, she discovered a complicated difference in how the environment shapes dendritic trees in males and females. When she repeated the experiment with males castrated at birth and intact males, she found the same pattern of dendritic growth in the brains of castrated male rats that she had documented in normal female rats. The hormones determined how the brains would react to the environment.

These results have interesting implications for people, Juraska says. When researchers discovered more than thirty years ago that environmental stimulation triggers growth in rats' brains, it 'had a very big impact on how we feel about humans,' she notes. Psychologists became convinced that it was vital to give children plenty of mental stimulation so as to induce neuron growth and help a child's brain reach its full potential. That's still true, but it may not be the whole story. If human brains respond to the environment in a way similar to the rat's brain, then environmental stimulation may elicit different types of growth, or growth in different parts of the brain, in boys and in girls.

Juraska's work could also provide the biological underpinnings to explain what many researchers have noticed about humans, and human children in particular: males and females respond differently to their surroundings.

They find different things interesting and they interact differently with the same environment.

It begins early in childhood, says Diane McGuinness of the University of South Florida. From the time they are a few months old, females pay more attention to and respond more to sounds, particularly voices and music, than males do. It's possible, she speculates, that the female advantage in verbal ability begins here – if females find speech more stimulating than males do, they may spend more time listening to it and stimulate the verbal areas of their brains more.

Sight, on the other hand, seems to be equally stimulating for both sexes, McGuinness says, but males and females differ in the types of things they respond to. All infants like to look at human faces, but from about four months on girls can distinguish facial features better than boys. Girls are better at telling one face from another and also at discriminating between a realistic and unrealistic line drawing of a face. This female advantage may lead to girls spending more time looking at faces since, being more sensitive to details, they get more stimulation out of them. Boys, on the other hand, are more interested than girls in objects – geometric patterns, three-dimensional shapes, blinking lights and colored photos of objects.

McGuinness found similar differences in preference among adults. She and John Symonds tested men and women with a tachistoscope, a device that simultaneously displays two pictures, one to each eye. If the pictures are shown very briefly, they compete for the brain's attention and the brain will perceive only the one it finds most interesting. McGuinness and Symonds tested subjects this way when one picture was of a person and the other of an object. When asked what they had seen, the women more often reported seeing people and the men were more likely to have noticed the objects.

McGuinness suggests that it is this sex difference, or one like it, that is behind at least some of the difference in the types of toys that boys and girls prefer. She thinks that girls like toys, such as dolls, that offer a 'human' stimulus, while boys are more likely to choose things that engage their spatial senses, such as cars, trucks and blocks. In particular, she says, male children prefer play that involves manipulation – moving an object with their hands and watching what it does.

Because of such differences in interests and preferences, boys and girls may learn very different things even when raised in identical surroundings. They 'make their own environments', in the words of Yale psychologist Sandra Scarr. Offered the same toys, they choose which toys to play with – and also how to play with them. A boy will sometimes play with dolls, for instance, but it's usually not to put them in a doll-house or set them down at a table for tea. More likely he'll place a doll atop a truck and send the truck crashing into a wall so he can watch the doll go flying.

I remember the last time I visited my brother and his family. I had taken the three oldest kids out in the front yard to play and so give their mom a little break. Six-year-old Richard loves baseball, and I pitched ball after ball for him to hit. Four-year-old Abigail chased after a ball or two but for the most part was content to stand and watch, even after I asked if she wanted a chance to hit the ball. Two-year-old Timothy, on the other hand, insisted on trying to throw the ball and swing the plastic bat, with little success but great enjoyment. (All three liked the game where Uncle Robert picks them up by their feet and swings them around and around until he's too dizzy to stand up any more.) Each of them had very distinct – and stereotypic – preferences as to the types of games they wanted to play. Yes, it's possible that even at four years old Abigail had already learned that baseball is for boys, but I don't think so. Swinging a bat at a ball just didn't seem like a lot of fun to her.

And it's not just their physical environment that children shape. They also take an active role in determining their social environment: interactions with mommy and daddy, playing with friends (and uncles), listening to teachers. Researchers have found that beginning shortly after birth children act in ways that trigger certain responses from others – and these environment-shaping behaviors often differ between boys and girls.

At one day old, there are few things that a baby can do to draw a response from an adult. It can cry. It can grasp a finger placed into its hand. It can establish eye contact, fixing its gaze on a person's face. And already, even with this limited repertoire, boys and girls are garnering slightly different reactions from adults.

In 1979, Joan Hittelman and Robert Dickes of the State University of New York Downstate Medical Center reported the results of a test on thirty newborns, fifteen male and fifteen female, who were between twenty-four and sixty hours of age. Since eye contact is perhaps the first social behavior that a baby can engage in, and since other researchers had found that mothers get pleasure from eye contact with their newborns, Hittelman and Dickes decided to look for a difference in how male and female babies gaze into a person's face.

After each newborn was fed and was fully awake, an experimenter tried to make eye contact with the baby in four positions – lying flat in a bassinet, cradled flat in the experimenter's arms, cradled in a partially upright position, and held fully upright. The number of times and amount of time that each baby looked into the experimenter's face was recorded over a period of about four minutes. When all the babies had been tested and their eye-contact times analyzed, Hittelman and Dickes found a large disparity. The females spent an average of fifty percent more time in eye contact with the experimenter than the males, and seven of the fifteen male babies fell below the lowest female score. The males didn't look at the experimenter's face any fewer times than the females, Hittelman and Dickes found, but the female babies maintained eye contact for longer periods.

The importance of the experiment, the researchers say, is that it shows male and female babies are already interacting differently with adults in the first few days of life – even if it is only because females have a slightly more mature visual system at birth. Since other research has indicated that face-to-face interaction between a parent and infant is a key to communication between them early in the child's life, boys and girls may elicit different behaviors from their parents almost from the time they are born.

This pattern continues through childhood. For instance, in watching young children with their mothers, Michael Lewis observed that the women gave more physical attention to their three-month-old sons than their daughters, touching and holding the boys more often while talking to or just looking at the girls more. This may not have stemmed from some unconscious bias on the part of the mothers, however. Lewis also found that the girls were more responsive to the mothers than the boys were, so the mothers of the boys may have had to work a little harder to get a response. And this in turn could have other effects on the child's behavior. Judith Rubenstein found, for instance, that children who get more physical comfort from their mothers are more likely to be adventurous in exploring away from mommy.

Perhaps the strongest way children shape their social environments is by their choice of friends, and beginning in the first few years of life, boys and girls prefer to hang around with members of their own sex. This appears to be innate – it's common among young primates to spend more time with same-sex peers, for instance, and children exposed to unusual hormone environments in the womb sometimes reverse the pattern.

Homosexuals of both sexes often remember preferring the company of the opposite sex as children, for instance, and in a recent study Sheri Berenbaum has reported that a significant portion of CAH girls were more like boys than other girls in their sex choice of playmates. Most of the CAH girls in her sample were similar to normal girls in their choices, but about ten percent seemed to prefer boys instead of other girls to play with, something that none of the normal girls in the study did. Sex segregation is universal among human cultures, and even in Western culture, with all its sexual stereotypes, there is little evidence that human parents push their children to play with one sex instead of the other. Kids do it naturally.

This is a different way to think about the contributions of nature and nurture to a child's development. Instead of passively absorbing whatever the environment has to offer, a child is actually an active participant, deciding (consciously or subconsciously) what to pay attention to and how to react, and even influencing how other people treat him or her. A number of researchers have suggested that sex differences in adults may have their roots in the different ways that boys and girls interact with their environments while growing up.

The male advantage in spatial ability, for instance, might derive in part from boys' preference for manipulating objects. Picking up blocks, turning them over to look at them from different angles, and assembling them into castles could trigger development in those parts of the brain that handle spatial skills, and there is some evidence that it does. In one study, Nora Newcombe from Temple University in Philadelphia tested the spatial skills of several dozen undergraduates and then compared their scores with their experience in a variety of spatial activities, everything from carpentry and making model airplanes to weaving and figure-skating. On average, the more spatial activities a student had engaged in, the higher his or her spatial ability was, and the correlation was particularly high for the female half of the sample.

The problem with such studies, however, is that it's difficult to tell whether a person has good spatial ability because he (or she) did plenty of activities that exercised that ability, or if the person chose those activities because he had good spatial ability to begin with. You like what you're good at. If your mind naturally handles spatial information with ease, you're more likely to enjoy playing with erector sets or sketching house plans. (This doesn't mean that spatial ability can't be improved with practice. Researchers have found that with special training it is possible to improve children's spatial skills, both boys' and girls'. But parents who want to ensure that their daughters have high spatial skills may need to do more than buy them building blocks and erector sets.)

In those children whose minds are primed to develop good spatial skills, the process probably depends on an interplay between nature and nurture. Kids with naturally good spatial ability are likely to play with things like blocks or Lincoln logs because they get pleasure from them, and in turn playing with those toys helps sharpen spatial skills. But like an infant with a cataract in one eye, a child who was born with a capacity for good spatial ability yet who never participated in spatial activities would never develop that talent.

And it is here that sex segregation may come into play. Since much of a child's play is done not alone but with other children, a child's tendency to choose playmates of the same sex will influence the types of play he or she engages in. Julia Sherman calls this the 'bent twig' hypothesis – 'As the twig is bent, so grows the tree' – and uses it to explain how a natural male advantage in spatial ability may be exaggerated by socialization. Males engage in activities as a group, and what the group does is determined by a sort of rough democracy. If many of the boys want to do something, they all do it, with even the dissidents willing to go along as the price for being with the group. Since most boys enjoy things that engage their spatial senses – playing with blocks, hitting a baseball, doing carpentry – even boys who might not have a predisposition for such activities get exposed to them. When I was a teenager, for instance, boys liked to work on cars, usually in groups of two, three or more, and the need to visualize how the pieces – the pistons and the rings and the camshafts – fit together certainly exercised your spatial skills. Of course, cars today seem too complicated and computerized for a bunch of teenaged boys to take apart and put back together, but I'm sure other things (video games?) have taken their place.

According to the bent twig hypothesis, a small innate sex difference can be magnified by this type of group behavior. If boys start out with slightly more enjoyment of spatial activities and a minor superiority in spatial skill, they can eventually develop a big advantage because they do things to practice this skill. With more skill they enjoy such activities even more, which makes them even more likely to choose spatial activities in the future, and so on in a self-reinforcing cycle. And society's tendency to stereotype the sexes can add to the snowballing effect – if groups of boys are more likely to work on cars than groups of girls, it becomes a 'boys' activity' and girls are discouraged from taking part.

At the very least, this type of natural sex-stereotyping of different activities can ensure that, for instance, most boys are exposed to video games whether they really enjoy them or not, while some girls who might enjoy them never find out. In *Sex Differences in Cognitive Abilities*, psychologist Diane Halpern writes that when her daughter registered for classes in the eighth grade she was told she had a choice between drafting or home economics. Each class was open to both sexes, but if her daughter signed up for drafting she could expect to be the only girl in the class. It was no contest, Halpern says – at an age when kids are especially sensitive to peer pressure, there was no way her daughter would choose drafting, even if that was the subject she would prefer.

Think of it as the 'stunted twig' hypothesis: if a child has desires or abilities that don't match with the norm for his or her sex, they're less likely to be developed because there is no positive feedback from peers. A girl with an aptitude for spatial activities may never develop it because none of her friends do.

Sex segregation is not an absolute rule, of course, and some girls do end up playing with boys (although few boys play with groups of girls). When my wife was little, she tells me, she played basketball and football with the neighbourhood boys, and they'd come to the front door and ask, 'Can Amy come out to play?' In fourth grade she was faster than any of the boys, and in sixth grade she won the free-throw shooting contest at her school (boys and girls).

But as children get older, they tend to spend much of their time in single-sex groups and, as Northwestern University's Janet Lever noted, this creates a difference in the types of skills they develop – especially social skills. Lever, who studied the play habits of fifth-graders, found that baseball and other boys's sports taught them to co-operate in large groups toward a common goal, to play according to fixed sets of rules, and to compete with other groups of males. Girls' games, on the other hand, prepared them more for interactions in small groups, where members are sensitive to each other's needs and keep open competition to a minimum. The different skills learned by the two sexes, Lever said, may well make a difference in how they handle jobs and social interaction as adults.

All this makes it impossible to separate nature from nurture when trying to figure out why boys and girls, men and women are different. Certainly the sexes are exposed to different environments when growing up, but those environments are at least partly of their own making. As

Janice Juraska demonstrated with her rats, even identical environments can stimulate males and females to develop in different ways.

One of the most common questions I hear from people who discover I'm writing a book about sex differences is, 'So how much is nature and how much is nurture?' I tell them that it's impossible to separate the two and I try to explain the give-and-take between biology and environment, perhaps by pointing out how boys and girls tend to spend time with members of their own sex and explaining how that can steer their socialization in different directions. But often I can see that this just isn't enough. The person wants a more practical answer because he's really asking a more practical question: if we treated males and females exactly alike, how much of the sex difference would still remain?

Even sex difference researchers disagree sharply on this, particularly when it comes to specific sex differences. How much of the male advantage in high school and college mathematics would disappear if we completely rid ourselves of the idea that math is a boys' subject? How much of the greater female nurturance would dissipate if boys were taught that taking care of children is not just a woman's job? Would males be so much more aggressive than females if boys and girls were equally discouraged from fighting? Would females lose their verbal advantage if parents talked to boys and girls just the same?

It's impossible to answer any of these questions exactly, of course, but the past ten years of sex difference research do point toward one broad conclusion. Most of the researchers I've spoken with would probably agree with this rather cautious statement: if boys and girls were raised in a world where parents, teachers and society responded to them as individuals, not as members of a particular sex, then some clear, sizable differences between the sexes would still remain. They would be smaller than they are now, but they would not disappear.

Since this is a rather touchy subject, let me make it clear right now that I'm talking only about sex differences in mental and psychological traits, not about such things as the number of men and women in different jobs, how much men and women get paid, and who does the housework and takes care of the kids. Such social and economic details are separate from the science and, to a large extent, independent of it. I'm also not considering the possibility that parents or society at large would treat boys and girls differently in a concerted effort to get rid of sex differences completely – for instance, by offering remedial spatial ability classes to girls only, or teaching boys to be more nurturing and less aggressive.

The evidence that there are natural differences between the sexes which would remain even in a non-sexist world comes from a number of directions. In Chapter 4 we heard what is perhaps the strongest testimony: the studies of people, such as CAH females, who were exposed to unusual hormone environments in the womb. The CAH girls, for instance, are raised just like normal girls, yet they prefer boys' toys to girls' toys and are often tomboys, and their scores on mental rotations test are nearly as high as males. Because there seems to be no way to explain this in terms of socialization, researchers like Sheri Berenbaum conclude that testosterone in the womb plays a major role in making boys' play patterns different from girls and also accounts for much of the male advantage in spatial ability.

The CAH studies don't reveal whether testosterone increases spatial ability directly, by altering brain circuits, or indirectly, perhaps by increasing the enjoyment a child gets out of manipulating objects, but other work implies it may be the former – that testosterone reconfigures a brain during development to give it a spatial boost. In testing CAH females, Berenbaum and colleagues found no correlation between their scores on tests of spatial ability and how much they engaged in childhood activities that might have exercised that ability.

And when two Boston College researchers, Beth Casey and Mary Brabeck, identified a group of female undergraduates who scored as high as males on a mental rotations test, they found that experience with spatial activities such as carpentry and building model planes did not seem to improve spatial ability by itself. Instead, they said, such spatial experience led to high spatial ability only in a subset of women who seemed to have a biological predisposition to develop it. This work combined with Berenbaum's implies that such activities as playing with erector sets and repairing car engines are probably not enough to develop high spatial ability by themselves, although they may help sharpen spatial skills in those with a natural talent.

In addition to the so-called 'experiments of nature' such as CAH girls, XY females and Turner's women, some evidence about human sex differences comes directly from nature itself – in the form of humankind's closest relatives, the monkeys and apes. These primates are genetically quite close to humans – the DNA of humans and chimpanzees, for instance, is more than ninety-eight percent identical – and the similarity is obvious not just in physical appearance (Who hasn't looked into the eyes of a chimpanzee in a zoo and felt a close kinship?) but also in various behaviors. And though the existence of these parallel behaviors is no proof that they're caused by the same factors, scientists assume that if humans act like monkeys in certain ways then there just may be a biological reason for it.

Rough-and-tumble play is the best example. In most or all monkeys, young males are much more rambunctious than young females, chasing each other around and play-fighting. The consistency of the pattern among monkeys and the close parallel with humans implies that testosterone in the womb is likely to blame for many six-year-old holy terrors.

Even in monkeys, however, testosterone is not the only factor in rough play – as in humans, the available playmates also have something to do with it. David Goldfoot and Deborah Neff observed juvenile rhesus monkeys living in groups of five. If the group was all female, the monkeys engaged in very little rough-and-tumble play. If there were two or more males in the group, the males tussled with one another and again the females took little part in the rough play. But if there was one male and four females, Goldfoot and Neff found that the females engaged in much more rough-and-tumble play. Apparently, it was the male monkeys who initiated the rough play, and if given the choice they would select other juvenile males to do it with, but if only females were available a male would try to pull them into his type of play and they would oblige him to some degree.

As adults, male monkeys and apes are generally much more aggressive than the females. Groups of male chimpanzees, for instance, will often kill a single male outsider if they catch him in their territory, and Jane Goodall has recorded one instance where the males of one group of chimps completely exterminated a second group, killing them one at a time when they caught them alone. Again, the consistency of the pattern and the correspondence to the human case hint strongly that the violence and aggressiveness of human males is more than just a product of socialization.

The flip side of aggressiveness in male monkeys is the fact that in most species of monkeys the females are more nurturing and take much greater responsibility for child-raising. Among the young of these species, 'play mothering' forms an important part of the play of females, but not males. Among vervet monkeys, for instance, it is the juvenile females who cuddle and carry small infants. Among year-old rhesus monkeys, females approach and touch infants three to five times as often as males. In fact, you can predict how much the father will be involved in taking care of the babies by watching sex differences in play mothering among the young – in the species where the fathers actually do some fathering, the juvenile males also show more nurturing behavior.

As was the case with rough-and-tumble play, child care in monkeys may vary with the social circumstances. In one experiment, an adult rhesus monkey was placed in one cage and a 'stranded' newborn monkey was put in an adjacent cage connected to the first by a passageway. Adult males were just as nurturing as adult females in the sense that they would cross to the other cage and pick up the young monkey as quickly as the females. But when both a male and female adult were put in the cage, it was always the female who would go take care of the baby. Even males who had comforted the youngster previously when no female was present now ignored it. (Those monkeys sound pretty human, yes? I must say, however, that from what I've seen many of the fathers in this generation do share responsibility for their kids and will pick up the baby or kiss a boo-boo even if mom's in the room. Maybe humans are starting to evolve past the monkeys.)

If humans follow the primate pattern, then it appears that females have a greater predisposition to mother small children than do males. One of the biggest differences between little boys and little girls is that girls like to play with dolls, a practice that closely resembles the play mothering seen in monkeys and apes. Is this nature or nurture? Undoubtedly parents are more likely to give dolls to little girls and to discourage boys from playing with them, but the evidence from CAH girls indicates that doll play is strongly influenced by hormones in the womb and that parents are merely reinforcing a sex difference that was already there. The idea is anathema to many feminists, but a number of scientists – including women scientists – take the idea quite seriously.

The difference in nurturance seems to be particularly strong with infants. Fathers are often quite good with children eighteen months old or older, but as sociologist Alice Rossi has pointed out, even in families where the parents split up the child-care duties, it seems to be the mother who is most nurturing to and gets the most enjoyment from infants in their first year.

None the less, the research also demonstrates that males can learn to take care of children and to get pleasure from it – a variety of studies on rats, monkeys and humans have shown that males' nurturing behavior to the young increases with exposure. Much more than spatial ability or even aggression, nurturance to children seems to be modified by the environment. The parenting push is probably stronger in women than in men, but that doesn't mean women are inevitably better parents than men.

Moving from primates back to humans, there is a somewhat indirect – but still very powerful – way to compare the contributions of nature and nurture to sex differences. We saw its tracks in Chapter 6, when Michael Bailey and Richard Pillard studied identical twins, fraternal twins and adoptive siblings to estimate how much of homosexuality is inherited. The field, behavioral genetics, compares related and unrelated people to get a fix on how much of human variation is due to biology and how much to environment.

'Everything is fifty percent genetic,' Bailey says. 'I call that Bailey's Rule.' He's referring to the fact that for a wide range of mental and psychological characteristics, behavioral geneticists have calculated that about fifty percent of the difference among people is due to their genes, the other half to the environment. Sometimes it's more, sometimes less, but if you don't know the actual number, just guess fifty percent and you're probably pretty close.

The evidence comes from a variety of investigations. The most powerful and best known are the so-called separated twin studies. Here a researcher tests sets of identical or fraternal twins who were given up for adoption to different homes while still infants. Because identical twins have the same genes, it's possible to estimate the relative influence of the genes by studying identical twins raised in different environments and comparing them with, for instance, fraternal twins raised apart or identical twins raised together. Behavioral geneticists also study adopted siblings to get a measure of the effect of a shared family environment on children with no shared genes.

The separated twin studies have turned up some truly spooky similarities, enough to warrant a cover on *The National Enquirer*. Take the case of two twins named Jim separated four weeks after birth. Both Jims, who were thirty-nine when they were reunited, were in their second marriages, each to a woman named Betty. Their first wives were each named Linda. They each had a son named James – James Allan and James Alan. Both said math was their favorite subject in school, each had a workshop in his basement for working with wood, and both liked watching stock car racing, smoking Salems and drinking Miller Lite.

Now, nobody is suggesting that there are 'name genes' – one for your kids' names, one for your spouse's name and yet another for your second spouse's name – but the other details may be more than coincidence. There probably is some genetic predisposition to smoking and drinking, and the genes for taste-buds should influence which brands taste best. Most researchers would accept the idea that genes have some control over a person's enjoyment of studying math or working with wood. And recent studies have found that even the likelihood of getting divorced is influenced by genes.

But past providing such head-scratching coincidences as both Jims having served as part-time deputies in their respective Ohio towns and both spending their vacations at the same small beach near St Petersburg, Florida, separated twins offer researchers the chance to compare objective measures, such as IQ or personality traits, in people who have the same genes but different upbringings. When Thomas Bouchard at the University of Minnesota tested the Jim twins, for instance, he found they scored so closely on many scales that they could have been the same person.

By accumulating data on dozens of sets of separated twins and combining it with information from other genetic studies, Bouchard and other researchers believe they can assign rough values to nature and nurture in many domains. The most heavily studied area is intelligence. The IQs of identical twins raised together correlate about eighty to ninety percent, according to several recent studies. (Correlation is a technical term that refers to how closely two sets of numbers match. If they match completely the correlation is one hundred percent; if they're no better than a random match the correlation is zero.) In other words, if you grew up with an identical twin, his or her IQ will normally be nearly identical to yours.

How much of this is due to shared genes and how much to shared environment? Studies of identical twins raised apart find a correlation of more than seventy percent – nearly as high as the correlation of identical twins raised together – so being raised in two different homes by two different sets of parents, going to two different sets of schools, having two different sets of friends, etc. doesn't cause much difference in how well two twins score on an IQ test. In numerical terms, researchers calculate that about three-quarters of the variation in intelligence among individuals raised in the United States or another Western country can be traced back to the genes.

Verbal and spatial abilities are less heritable than IQ, but still strongly dependent on genes. One recent study of separated twins in Sweden found that nearly sixty percent of the variation of verbal ability among individuals was due to genes, as was almost half of spatial ability, and also about sixty percent of perceptual speed. Memory seems to have a smaller genetic component, as its heritability was calculated at less than forty percent.

Personality traits are generally less biological and more environmental than cognitive abilities, with heritabilities of about forty to fifty percent, and the most heritable traits seem to be sociability, emotionality and activity level. A recent twin study by researchers at the University of Southern California put the heritability of masculinity at forty to fifty percent and of femininity somewhat less, twenty to thirty percent. (In this study, masculinity meant having such traits as dominance and assertiveness, while femininity boiled down to nurturance, interpersonal warmth and other such expressive traits. Psychological researchers define masculinity and femininity in various ways, and the terms often refer to little more than those psychological traits that are statistically more likely to appear in men than women or vice versa.)

To many people these numbers are surprising, even shocking. In Western society it has long been taken for granted that much of the variation in people's mental and psychological traits could be traced to differences in their environments, but now the evidence says that most of the variation in intelligence is due to genes. Even something as personal as whether you're an extrovert or introvert is fully half genetic.

And that's not all. The other twenty-five or fifty percent of variation due to environmental factors does not arise from what most people think of as the most important environment – the atmosphere that exists in a home and is shared by brothers and sisters. By comparing twins raised together and twins raised apart, or unrelated children (adopted siblings) raised in the same home with unrelated people raised in different homes, researchers can get a measure of what they call 'shared environmental factors'. These are the influences that siblings in the same house share – the intellectual stimulation provided by the parents, the values taught in the home, physical factors such as nutrition, and so on.

The research is unanimous on this: shared environmental factors contribute almost nothing to a person's mental make-up. The IQs of unrelated adoptive siblings raised in the same home correlate about two percent – they aren't much closer than the IQs of perfect strangers. For personality factors, the correlations between biologically unrelated siblings average about five percent.

It is true that early in life children show the influence of the home they're raised in, but those effects evaporate as they get older. The Texas Adoption Project, which has been following several hundred adopted children through childhood and into adult life, found that when they were young, adoptive siblings had IQs closer than expected, given that they were not genetically related. But in a ten-year follow-up, when the adoptees averaged eighteen years of age, their IQs had almost zero correlation with those of their siblings. Parental influence on a child's scores on intelligence and personality tests is strongest in childhood and retreats almost completely by the time the children are adults themselves. Whatever the environmental factors are that influence a person's mental or emotional characteristics, they are mostly 'unshared' – different even for siblings living in the same house.

Researchers are still debating what these unshared factors are. They probably include influences from outside the home (friends, teachers, Saturday morning cartoons) as well as in-home experiences that differ from child to child. Parents do not treat children exactly the same way, of course. The firstborn, for example, always seems to be handled much differently by parents than the siblings that follow, once mom and dad are old hands at the parenting game, and growing up as the firstborn, with the house to yourself for the first years of life and no older siblings to learn from, is a much different experience than growing up as a middle child or the baby. There may also be some non-genetic biological influences, Bailey says. The womb environment that a mother provides is different from child to child, and even twins may have different fetal environments based on their position in the womb. These factors could affect brain development.

None of the behavioral genetics results apply directly to the question of nature/nurture in sex differences, but they do offer some general guidelines. They tell us that among a group of people growing up in relatively similar environments – middle-class Americans, for instance – genes account for half or more of the variation in how well they do on tests of verbal and spatial ability, while environment explains less than half of the differences among them. If you were to look at people from very different environments – Americans and Japanese, for instance, or children of wealthy parents versus kids growing up in single-parent homes in the inner cities – then you would expect the differences to be greater and environment to play a bigger part in creating those differences, but even then genes would still be behind much of the variation. The lesson for sex differences is that in people of roughly similar backgrounds, biology takes the lead in creating differences in such traits as writing a good paragraph or drafting an architectural blueprint, while environment has only a supporting role. Even if one child grows up in a house with a well-stocked library and another in a home where *TV Guide* is the main reading material, by the time they're adults the difference in their verbal skills will be more a matter of their biology than their upbringing.

It's still possible – and likely – that some percentage of the sex difference in verbal or spatial skills is caused by males and females being treated differently as they grow up. But the lesson from behavioral genetics is that as long as people are raised in relatively similar environments, most of the variation in their mental traits and much of the variation in their psychological traits is due to nature. In order for nurture to be responsible for a big difference among individuals, there must be some major differences in the environments in which they are raised. Such things as disparities in parental encouragement or having different toys to play with just won't have major consequences on average (although they might have big effects on some individuals).

So it's important to know just how differently boys and girls are raised. Are the differences between their environments so great that it's like growing up in two totally different countries, say Sweden and Sudan? Or are the dissimilarities no greater than growing up with the Smiths versus growing up with the Joneses down the street?

Over the past four decades hundreds of scientists have produced reports on how parents treat their sons and daughters, on topics ranging from the amount of discipline each sex gets to how much encouragement they receive on schoolwork and what types of things parents reward their children for. It's a Herculean task to make sense out of all of it, but fortunately I don't have to do it. Two researchers at the University of Calgary in Alberta have already done it for me.

In 1991, Hugh Lytton and David Romney published an article surveying one hundred and seventy-two papers dating back to 1952 which studied sex differences in socialization in North America, Europe and Australia. They used a statistical technique called meta-analysis to extract an overall meaning from this great mass of data, and what they found casts doubt on the idea that girls and boys grow up in quite different environments.

There is one way, and only one way, Lytton and Romney concluded, in which parents do treat their sons and daughters very differently. Parents teach their children how males and females are supposed to act, according to society's notion of proper sex roles. In the United States this means the little girls are put in dresses, allowed to grow their hair long and given make-up, but boys are not. Little boys are encouraged to be tough, not to cry, to be leaders. Girls aren't. Daughters help with cooking, sewing, cleaning house. Sons take out the garbage and help with the yardwork. Boys are encouraged to play male games, especially sports, and are more likely to get boy toys as gifts – cars, trucks, construction sets, sports equipment. Girls are given dolls, tea sets, kitchen toys. Girls grow up expecting to get married and raise a family, while boys learn that they're supposed to get a job and pursue a career. These differences were true in the 1950s and, although they may be diminishing, they're still largely true today.

But when Lytton and Romney looked for differences besides the encouragement of proper sex roles, they found very little. In North America, sons are more likely to be spanked or receive other physical punishment than daughters, but not by much. Boys get more prompting to achieve, are disciplined more strictly and are subject to more restrictions, while girls are treated more warmly and are encouraged to be more dependent, but all of these differences are small. In each category, some studies will find a difference, others won't, and some will even report a disparity in the opposite direction. For instance, some researchers reported that dependency was actually encouraged more in boys than girls, while others found that girls were pushed more to achieve than boys. The pattern of results was far from consistent. Despite the stereotypes, Lytton and Romney concluded, parents in Western societies don't treat their sons and daughters that differently, except to teach them how society expects boys and girls to behave.

Is that enough to create the observed sex differences in cognitive abilities and psychology? Many sex difference researchers think not, given that all environmental influences, sex-biased or not, account for less than half of the variation in cognitive abilities among people in a population. 'In view of all the evidence,' Lytton and Romney wrote rather cautiously, 'we cannot close our eyes to the possibility of biological predispositions providing a part of the explanation for existing sex differences.'

One weakness of much of the research on sex differences is that, like Lytton and Romney's work, it focuses only on Western societies. There are few studies of sex differences in Eastern countries, such as Japan, China or Korea, and fewer still that are performed in Third World countries, where the cultures are much different. Do women have better spatial ability than men in Bangladesh? Do men outscore women on mental rotations in Botswana? We don't know.

Yet it is particularly helpful to compare men and women in different cultures. We know, for instance, that there are very violent peoples, such as the Yanomamo of Venezuela, and very peaceful peoples, such as the Semai of Malaysia. Do sex differences in aggression exist in either of these tribes? If not, then we should ask what it is about our culture that brings out such a difference. And if anthropologists were ever to discover a group of people in whom the women are clearly more aggressive than the men, then we would suspect that sex differences in aggression are created by how males and females are raised, not because of biology.

This is the basic idea behind cross-cultural studies. If a certain human behavior is culturally determined, it should differ from one society to the next. To take a rather silly example, suppose somebody told you that there is a biological reason why women wear dresses and men wear pants. It's a natural, hormone-driven sex difference, this person says. To test this claim, all you need to do is to gather data on a number of human societies from around the world. When you find that men wear skirt-like clothing in a number of cultures, you can discount the hormone-clothing connection.

Conversely, if you find that a particular behavior is universal, or nearly so, across cultures, you can be pretty sure that there is some biological predisposition for it. All human cultures have spoken languages, for instance, which means that speaking is something humans do quite naturally. On the other hand, not all cultures have written languages, implying that writing isn't nearly so natural.

How natural are sex differences? We can get a good idea by looking for patterns across cultures.

The most consistent cross-cultural evidence concerns aggressiveness. In all cultures known to anthropologists, the male is more violent than the female. In the United States and other Western societies, men are responsible for the lion's share of the homicides, with about eighty-six to eighty-eight percent of those arrested and charged with homicide in the United States being men. And when a woman does kill somebody, it's almost always a husband or a boyfriend, not a stranger. Nor is the male's near-monopoly on murder just a Western phenomenon. Consider the peaceful !Kung San of the Kalahari Desert, a tribe of hunter-gatherers who live much as anthropologists believe all humans lived ten or twenty thousand years ago. (The '!' is a phonological symbol denoting a clicking sound that has no equivalent in English.) Among the !Kung San, a tribe known for its sexual equality, every one of the twenty-two homicides documented by one researcher was done by men – and all but three of the victims were men. Some societies are more violent and aggressive than others, but within each society the pattern remains the same: men are the more savage sex.

Unanimity of this sort can only mean that a biological mechanism is at work. It's possible to debate whether a given society exaggerates or dampens the natural difference in aggression between the sexes, but a difference there is.

Other social sex differences are not as pronounced across cultures, but a few are consistent. Perhaps the best cross-cultural study of sex differences is the 1973 report by Beatrice Whiting and Carolyn Pope Edwards on behavior in children aged three to eleven in six countries around the world: Kenya, Okinawa, India, the Philippines, Mexico and the United States. Besides finding that boys were physically and verbally more aggressive around the world, Whiting and Pope saw several other reliable patterns.

Girls were more 'dependent' than boys. The younger girls, three to six, were more likely to ask for help than boys, although the sex difference disappeared as they got older. And all of the girls were more eager for physical contact, touching, holding and clinging to others more often than the boys. The boys, however, had their own form of 'dependency', Whiting and Edwards noted. They sought attention and approval more, especially as they got older, trying to get either a positive or negative reaction from people nearby. In short, males and females have different ways of interacting with other people, differences which are apparent at an early age and which are consistent across very different cultures.

The other major difference was that girls were more nurturing to other people than boys. They consistently scored higher both on offering help, such as food, tools or toys, and on offering emotional support and comfort. The difference increased as the children got older, and Whiting and Edwards attributed the trend to the societies training girls to be more nurturing toward infants in preparation for motherhood. As evidence the researchers pointed to two exceptions to the rule, societies where boys and girls were equal in offering help and support. Both exceptions were cultures that didn't expect girls to participate more than boys in caring for infants – Kenya, where young boys shared in helping care for infants, and a small town in New England, where families were small and the girls did little babysitting.

Whiting and Edwards conclude that sex differences in aggression and touching behavior are most likely to arise from biology. The sex difference in nurturance, they say, is at the very least exaggerated by cultural expectations and may be mostly due to young girls being expected to help with children.

Concerning sex differences in mental abilities, there have been few cross-cultural studies that targeted people outside North America and Western Europe because such research is expensive and time-consuming, but one recent test did compare sex differences in the United States and in Japan, a decidedly non-Western culture. Working with three collaborators from Japan, Virginia Mann at the University of California at Irvine created a battery of tests in English and Japanese and gave them to high school students in the two countries. In both countries males outscored females on mental rotations, while females outperformed males on word fluency (coming up with words that start with a given letter or, in Japanese, kana character) as well as story recall and digit-symbol, a common test of memory.

This research clearly demonstrates that nurture can play a large role in how people perform on tests. The Japanese students – who are pushed to excel by parents and teachers – consistently outperformed the Americans, and the difference was so large on the mental rotations test that Japanese girls outscored American boys. On the other hand, nature was just as clearly playing a role in creating differences within each culture. Despite the fact that children are raised quite differently in the United States and Japan, the students from these two cultures had very similar patterns of sex differences in their cognitive abilities. It would be nice to have more cross-cultural studies like this, with students from Paraguay or Pakistan or Pago Pago, but even without them, we can clearly see that similar sex differences appear in quite disparate environments – an indication of nature at work.

What differences would remain between boys and girls, men and women if our society were sex-blind, so to speak – if people were treated as individuals, not as members of one sex or the other? Boys would still be boys, with their rough-and-tumble play and their fascination with manipulating objects, from playing with blocks and trucks to building model airplanes and competing on video games. Girls would still be more people-oriented than object-oriented and would still like playing with dolls. The psychological differences that Carol Gilligan and Deborah Tannen have identified – females paying more attention to people's needs and to the web of relationships and males more attuned to rules of right and wrong and to dominance hierarchies – might still hold, assuming that they are somehow connected with the person/object difference in interests between females and males.

Females would retain much or all of their verbal superiority, and males would still have an edge on spatial skills. Some of the male advantage in higher-level mathematics would likely fade away, since it seems to have been slowly decreasing over the past couple of decades anyway, but boys would probably retain much of their lead among the very top scorers on math tests. Males would continue to be more variable in intelligence than females.

Men would still be more aggressive and commit more violent crimes than women. Women would still be more nurturing to infants, although it's possible that in most other ways males and females would be similar in nurturance. The evidence implies that much of helping behavior among both males and females is taught, and often girls are taught more of it than boys.

All this may make it seem as if the world would not be much changed if the sexes were treated equally, but I don't mean to imply that at all. Although these statistical sex differences would remain, many if not all would probably be smaller. And the consequences might be quite large for some individuals. There may be many girls, for instance, with a knack for spatial ability and mathematics who never pursue these interests and never develop their talents because they're intimidated by venturing into a 'male' field.

And I have said nothing about how much of the differences between the sexes might be erased by active intervention. We know that spatial ability can be improved by training, so the sex difference in spatial skills might be decreased if both boys and girls were given this training. The same is likely true for other areas. In a society that taught non-violence to all its children, males might still be more aggressive than females, but the difference might well be smaller.

In general, I've written little about the massive amount of research that has been done on how environment and socialization affects boys and girls. That is not because it's unimportant, but simply because it is another book. The question here was what type of role nature plays in creating differences between males and females. The answer is: a major one.

CHAPTER 10

Echoes of the Past

There are whys, and then there are whys.

Why are men and women different? Scientists like Janice Juraska, Sheri Berenbaum, Elizabeth Hampson, Laura Allen, Christina Williams and the others we've met offer one sort of answer. Their whys speak of sex differences in the brain, of the way that hormones push a mind toward the male or the female. It is an approach that an engineer could love: how is the brain designed? And how is that design different in men and women?

But there is another sort of why, a why that goes deeper than how and asks what it all means, a why for philosophers instead of engineers. Is there some reason, some purpose, for men and women being two such different creatures? What do sex differences tell us about the heart and soul of humankind?

These questions too have answers – albeit more speculative – which have become clearer over the past few years. True, the answers aren't very big on cosmic significance or metaphysical truths. If that's what you seek, you still have to look to the Bible, the Koran, or other religious writings of your choice. But if you'd simply like a little better idea about why we are who we are, about what led to men and women being such different creatures, then read on.

My own search for why led to Steve Gaulin, an evolutionary biologist at the University of Pittsburgh. Gaulin has spent much of his career studying the evolution of sex differences, and over the past decade he has performed a series of experiments that reveal why a sex difference in spatial ability should develop in some species but not in others. So one winter afternoon in Pittsburgh we sat at his dining-room table and talked about the forces that make male and female different.

Our interview began slowly, unhurriedly, with Gaulin telling me a bit about his life – how he dropped out of high school at sixteen to get married, had two kids by eighteen and supported the family by working as a surveyor for an engineering company. Wanting something more out of life, he got a GED (graduation-equivalence degree) and entered Long Beach State College at twenty-one. After two years he transferred to the University of California at Berkeley to obtain a BA in anthropology and psychology, and he capped off his education at Harvard, earning a Ph.D. in anthropology with a dissertation on the feeding behavior of howler monkeys. Now, at forty-five, he's a full professor at the University of Pittsburgh, a respected evolutionary researcher, and a grandfather.

After the biography, we touched briefly on a couple of studies on sex differences he had been working on. We talked about some land he'd just bought in Arizona. And periodically we looked out the window at the four bird feeders in his yard and their steady stream of clients – sparrows, juncos, titmice, chickadees, goldfinches, wrens and a spectacularly colored pair of red-bellied woodpeckers. Gaulin, an avid birdwatcher, instructed me in the finer points of identifying birds. He pointed out the flash of white a junco shows as it flies away and told me I'd come too early to see the goldfinches at their best. A drab mustard brown then (early February), they would be brilliant yellow in a month, once the mating season started – but only the males. Like many other birds, the goldfinches have a very noticeable sex difference in their plumage and, like most of the species with a sex difference, they follow the 'Florida Retirees Rule': it's the males who get to wear the colorful jackets.

But Gaulin, as much as he loves birds, made his name by working on a very different critter. So, leaving the bird feeders, we begin to talk about spatial ability, sex differences and the vole.

The vole is a small rodent that looks a lot like a stocky, short-legged, short-tailed mouse. Gaulin chose to study voles because the various species have a range of mating patterns, some of which involve males staking out larger territories than the females and some of which don't. Assuming that spatial ability would be related to size of the territory, Gaulin hypothesized that a sex difference in spatial ability would exist in those species in which the males covered more ground than the female but not in those where males and females had equal ranges.

To test his hypothesis, Gaulin settled on two common species of voles. The meadow vole, often called the 'field mouse', is a mostly solitary creature. Both sexes keep to themselves, staking out their own ranges in a field and trying to keep other voles of the same sex out of those territories. Typically a male claims a large range that will include the ranges of several females, and he mates with those females. The second species, the pine vole, lives in groups that include one breeding female, one or more fertile males, and various members of an extended family. All of the members of a group, male and female, have the same range. If Gaulin's hypothesis were correct, a sex difference in spatial ability would exist in meadow voles, but not in pine voles.

Working with colleague Randy FitzGerald, Gaulin first studied populations of the two types of voles in the wild. They trapped twenty or so of each species and outfitted them with miniature transmitters to keep track of how far they ranged. Their data agreed with what other researchers had reported: the meadow vole showed a big sex difference in the size of the range, while the pine vole showed none.

Then Gaulin and FitzGerald trapped the voles again and brought them back to the laboratory to test their spatial ability. They used a maze specifically designed to require the same type of directional sense the voles would need in the wild to wander around looking for food and still know which way was home. A vole would first be trained to follow a zigzag path from a central area to a finish point that had a bit of food. Then the maze was modified so that when the vole came into the central area it was faced with eighteen arms pointing out at various angles, one of them pointing directly at the place where the food had been before. Since the original path to the food was blocked off, the vole's task was to figure out which of the eighteen new paths pointed in the right direction and to head down that arm. A vole's performance was scored on how close it came to choosing the correct direction in two tries.

When Gaulin compared the maze performance by sex and by species, he found exactly what he had predicted: among meadow voles, where the male has the larger range, the males had a better sense of direction than the females. Among the pine voles there was no difference between the sexes. Since then, Gaulin and FitzGerald have repeated the experiment with a different type of maze and with prairie voles, another monogamous species, in place of pine voles. They got the same results – in the species where the male covers a larger range than the female, the males perform better in the laboratory maze, while in the other species there is no sex difference.

The sex difference appears to be innate, Gaulin says, and not caused by the males in some species getting more navigational practice than the females because they cover more ground. He has compared how voles caught in the wild perform in mazes compared with lab-raised voles. The voles that have been kept in cages all their lives do just as well as their wild relatives, so running around a natural habitat doesn't seem to improve a vole's performance in a lab maze.

Furthermore, one of Gaulin's associates, Lucy Jacobs, has found a physical sex difference in the brains of voles that seems to correlate closely with the sex difference in spatial ability. It lies in the hippocampus, a small section of the brain that plays a critical role in spatial learning in rodents, birds and other animals. 'If you rub the hippocampus out, an animal can't find its way out of a paper bag,' Gaulin says, referring to a number of tests by other researchers on the role of the hippocampus. Jacobs, working with Gaulin, David Sherry and Gloria Hoffman, found that the hippocampus was larger in male voles than in females – but only in those species with a sex difference in spatial ability. In voles where the male and female had equally good directional sense, the hippocampus was the same size.

The message from Gaulin's vole experiment: when a species develops a sex difference, it is because the males and females have different needs. In this case, the male voles of some species (those in which the males try to mate with more than one female) need to be able to find their way around larger areas than the females do, and so those species have evolved with the males having better spatial ability. Other species have no need for such a sex difference since the males and females cover the same size territories, and sure enough, in those species no sex difference exists.

The general idea is not new, Gaulin notes. Darwin himself noticed that in some species one sex is larger or more colorful than the other, or perhaps has some feature – antlers or mane or ornate tail – that the other sex doesn't. Most of the obvious physical sex differences are related to reproduction – and in particular, reproductive competition. In many species of birds and fish, members of one sex (usually male but sometimes female) vie with each other to attract the opposite sex, which in turn is more choosy about picking a mate with just the right look. The more competitive sex often evolves eye-catching adornment which ranges from merely colorful, such as the goldfinch's jacket, to flamboyant, such as the tail of the peacock. Whenever you notice a species in which one sex is much brighter or has some fancy decoration, from songbirds to tropical fish, you can be sure that it reflects a sex difference in how males and females select or compete for mates.

In many mammals the competition is much more direct, with males fighting one another for control of either a breeding area or a group of females. This competition leads to a different sort of sex difference than the decorative ones in birds – male mammals are generally larger and more aggressive than the females and they may carry weapons to help in their combat. Male baboons, for example, have much larger canine teeth than the females. Male deer carry antlers. And male elephant seals, which engage in bloody combat to control stretches of breeding beach with dozens or hundreds of females, are three times larger than the females and have huge elephant-like noses they use in their fights. The two sexes look like different species.

Such sex differences, Gaulin explains, develop as part of the normal evolution of a species, but with a twist. A species evolves because its individual members have slightly different traits, some of which are more 'helpful' than others in the sense that they increase an individual's chances of surviving and passing on his or her genes to following generations. Over time, the genes for the more favorable traits will spread throughout the species while the genes for the less favorable traits will be weeded out. Random mutations in the genes allow new traits to appear, to be either kept or discarded depending on whether they are useful for survival and reproduction.

The same process, however, cannot by itself explain how a sex difference develops in a species. If a male vole has a gene for high spatial ability or a male goldfinch has a gene for a brighter yellow plumage, he will pass that gene on to all his offspring, both male and female. (Unless, of course, the gene is on the Y chromosome. But as we saw in Chapter 3, except for the 'maleness switch' and two other genes whose functions are not known, researchers have found no genes on the Y that are not also on the X, and so it's unlikely that sex differences derive from genes on the Y chromosome. That's true in humans, anyway, and most mammals seem to have a similar arrangement.)

This is where sex hormones come in. Since males and females are exposed to very different levels of hormones in the womb and later in life, a simple trick allows a species to evolve different characteristics in the two sexes: let the action of certain genes be influenced by sex hormones. If there is plenty of testosterone or some other male hormone around, a hormone-sensitive gene will do one thing; if there is little testosterone or perhaps a lot of estrogen, the gene does something else. This is probably what happens in the meadow vole, Gaulin notes. Male and female voles likely have the same genes to specify the development of the brain, but the male ends up with extra spatial ability because male hormones push certain of those genes to structure things differently in male brains.

This is the key to the evolution of sex differences, Gaulin says: the existence of genes whose effects are modified by the level of sex hormones. In this way a trait – the goldfinch's jacket, the elephant seal's size, the meadow vole's spatial ability – can evolve somewhat independently in the two sexes.

Gaulin's discovery in voles demonstrated that animals can develop sex differences in mental traits – in this case, spatial ability – for much the same sorts of reasons that they develop sex differences in physical characteristics. That is, males and females tend to evolve in separate directions when different traits are useful in the two sexes. And this, most researchers believe, is precisely what happened in humans. Because different qualities were useful to our male and female ancestors, over the course of thousands of generations the human race came to distribute some characteristics unevenly between the sexes.

This probably accounts for the human sex difference in spatial ability, for example. A number of researchers have suggested that the roots of this difference lie in a sexual division of labor among our ancestors. From the time that modern humans appeared, perhaps one hundred thousand years ago, to the time when agriculture was invented, about ten thousand years ago, humans lived in small groups of hunter-gatherers. To judge from the few hunter-gatherer groups left in the modern world, men and women in such cultures shared responsibility for providing food, the men by hunting animals and the women by gathering roots, berries, fruits, nuts and greens. No one knows how or why this sexual division of labor got started, but once it was established, the two types of jobs would have required very different skills. And, notes psychologist Irwin Silverman of York University in Ontario, those different sets of skills are very much like those we see in men and women today.

As hunters, early men probably threw rocks at their prey and later learned to make spears (bows and arrows appeared much later), but in either case males needed to be able to pitch something fast and accurately. And indeed, the male body today is much better designed for throwing than the female body. At eleven years of age, for example, there is very little difference between males and females in such athletic measures as grip strength, long jump and shuttle run, but the sex difference in how fast kids can throw a ball is huge – its effect size is about

$d=3.5$. (Remember that effect size indicates how big a sex difference is. The effect size for adult height, a relatively large sex difference, is about 2.6.) An effect size of 3.5 means that in an average group of one hundred eleven-year-olds, half boys and half girls, the fifty with the fastest throw would include forty-eight boys and only two girls. The fact that such a huge sex difference exists even before puberty implies that the male body evolved to throw things hard and fast.

But of course throwing something fast wouldn't do those paleolithic hunters any good unless they could also throw it accurately – at both stationary and moving targets – and so, Silverman suggests, over many generations males gradually honed their ability to hit a target. Or, to put it more accurately, males with better throwing skills were better providers and thus more likely to reproduce and pass on their genes to succeeding generations of males, leading human males to develop an aiming ability dependent upon their brains being exposed to testosterone. Today, dynamic spatial skill – the ability to predict the path of a moving object – shows one of the largest and most consistent sex differences of any cognitive ability. The advantage that males have at trap-shooting and shoot-em-up video games can be traced back to 'man the hunter', Silverman says.

Hunting probably also led males to cultivate a second type of spatial ability, he says. Hunters must be able to follow their prey for long distances and still find their way back, so part of the human sex difference in spatial ability likely arose for the same reason it did in some voles – the males had to cover more territory than the females. If Silverman is right (and other researchers, such as Doreen Kimura, have suggested the same thing), then the modern male's advantage on map reading and mental rotations tests can be traced back to his great-great-.... -great-grandfather's need to run up hill and down dale, climb trees, ford streams and still be able to visualize which way was home, sweet home.

Meanwhile, the men's paleolithic partners had their own duties. Besides having primary responsibility for child care (including breast-feeding each child for two or three years), a woman might walk many miles carrying one or two children in search of food to bring back to the camp. Studies of some twentieth-century hunter-gatherer societies indicate that, contrary to the stereotype of the hunters providing the lion's share of the diet, the food gathered by the women may have accounted for sixty to seventy percent of what the group ate, including more than half of the protein.

These gathering duties demanded their own specialized skills, Silverman hypothesizes. The women needed to locate plants with ripe fruits and nuts, and then remember where they were and find them again the next year. Since there might be only one tree or bush with suitable food in a whole confusion of other plants, one important skill would be a memory for how objects were arrayed with respect to one another.

Working with his student Marion Eals, Silverman designed several ways to test the particular type of 'spatial memory' – remembering things in their proper locations – that they thought would be important in gathering. In one test, for example, a subject was put in a room that looked like a typical graduate student's office, ostensibly to wait until the experiment started. But after a couple of minutes the subject was taken from the room and asked to name as many objects in that room as possible and give their locations. Women remembered seventy percent more objects than men.

In another experiment, the researchers first showed the subjects a piece of paper with a couple of dozen figures drawn on it – a hat, a bird, an ironing board, a flower and so on. Then giving the subjects a second page with the same objects plus a number of other drawings added, they asked the subjects to cross out all of the items that had not appeared in the original drawing. Finally, the subjects were handed a third sheet with the same items as the first but with some of them out of position, and this time they were asked to cross out the items that had been moved. Women scored about fifteen percent higher on both tasks.

Silverman and Eals believe that the women's superiority in these memory tasks is descended from the skills females developed tens of thousands of years ago in order to spot a likely food source and remember where it was in order to find it later. Women not only have a better memory for objects and their locations, but they naturally pay more attention to their surroundings. It is this tendency, Silverman and Eals suggest, that may explain why it's the wife, not the husband, who always knows where things are around the house, even if it's something the husband himself put down somewhere and can't find.

Doreen Kimura has argued that women's better perceptual speed (noticing details quickly) may have arisen from the advantage it gave them in spotting ripe fruits and other foods. It might also have been important in noticing small changes in an infant's appearance, she says. And because women probably had responsibility for such 'housekeeping' chores as making clothes and preparing food, it would have been more important for them than for men to have good muscle control for close-in activities, Kimura suggests. This may explain why women have better dexterity and fine muscle control, while men have better co-ordination for large movements, such as throwing a ball.

The female advantage in verbal ability has proved much harder to explain, however. Why should women who were responsible for taking care of children and finding food need better language skills than men going off to hunt? Some researchers have suggested that the mother–child relationship pushed women to speak more fluently since verbal skills would have been important in communicating with and commanding small children. Silverman and Eals suggest instead that the language advantage may have arisen as a way to help women recall where the food sources were. It would be easier to remember locations if the women had the words to describe those locations precisely to themselves and others.

On the other hand, Kimura hypothesizes that the female superiority in verbal fluency and other verbal skills may be merely a by-product of their superiority in fine muscle control. Noting that control of speech is handled in approximately the same part of the brain as the higher level co-ordination of muscle movement, she suggests that women first developed better fine motor skills and those skills gave them a built-in advantage in speech once early humans began to talk.

The reason for females having better verbal skills could be any one of those three, a combination of them, or perhaps another factor or factors that no one has thought of yet. It's difficult to say because we have only a general idea of what early humans were like and how they lived. We don't know how humans used speech fifty thousand years ago, or even when humans first began to speak.

And this brings up another problem in trying to understand the human mind now in terms of evolutionary pressures on humans many millennia ago. Our brains may have certain predispositions that are quite similar to those of our ancestors fifty thousand years ago, but those predispositions are certain to look very different in the light of today's culture. The human mind is an amazingly versatile, trainable instrument that can do countless things it was never 'designed' to do by evolution. Play the piano? Paleolithic man never did. Calculus? Not something that paleolithic woman taught her kids. Yet if you took a time machine back fifty thousand years ago, snatched an infant and brought him back with you to the present, he'd grow up just like any other child of the nineties. He could play the piano if he took lessons. He'd have as good a chance of passing calculus as any of his peers. And he'd like to listen to whatever awful music is popular with kids these days.

   The importance of culture and learning separates us from the voles and makes us much harder to understand. Voles are simple creatures. They have a certain type of spatial ability that helps them find their way around a field or maze, but it's not very flexible – they could never apply it, for instance, to solve a paper-and-pencil maze or read a map. Humans, on the other hand, have a generalized skill called spatial ability. We can use it for aiming a spear or visualizing how to get back from a hunt, but we can also use it to read blueprints, play video games or take mental rotations tests – things that weren't even imagined for most of human history when that generalized spatial ability was evolving. The sex differences we see today are products of the interplay between predispositions developed in men and women long ago and the environment that humans are raised in now.

   Take the sex difference in reading ability. Why should males have a more difficult time learning to read than females, and be three times as likely to be severely dyslexic? Yes, it's probably related to boys' general weakness in verbal skills compared with girls, but there's more to it. Reading is a relatively new skill, a skill the brain was not 'meant' to do in the sense that there was no evolutionary push specifically toward a brain that could read. It just so happens that the circuits the brain uses to make connections between written symbols and the spoken word are a little more suited to the task in females than in males. There's no good reason for it. It just happened.

   A similar thing is true for mathematics. We know that males and females tend to take different approaches to solving math problems and that there are more boys than girls among the very top mathematics students. It's certainly not because paleolithic boys with straight As in trigonometry were more likely to get a good job, get married, have lots of kids and pass their genes on to the next generation. Instead, researchers think that part of the sex difference can be chalked up to the difference in spatial ability, with boys being more likely to visualize problems spatially than girls, but there are probably other factors we don't understand yet. The bottom line is that males' and females' brains develop in slightly different directions because of things that were important tens of thousands of years ago, with the result that the male brain happens to be somewhat more predisposed to learning math, at least the way it's taught in schools today.

That's the basic pattern for explaining why sex differences exist: over the course of human history, males and females developed different characteristics because different traits were important to men's and women's success in surviving and reproducing. Those different predispositions are still with us today, but they develop somewhat differently in twentieth-century culture than they did in the world of 50,000 BC.

There's relatively broad agreement on this general pattern, but the details are still up for grabs. Gaulin, for instance, offers his own version of why males developed higher spatial ability than females. Like Silverman and Kimura he thinks it resulted from males covering more territory than females, but he has a different idea as to why they needed to cover more territory. The distinction between hunting and gathering isn't satisfying, he says, because women in their search for food sources might have covered just as much territory as the men did in their hunting. 'The meadow voles have a sex difference in spatial ability without a division of labor,' he notes, and he proposes that the human sex difference arose as it did in meadow voles – because of a difference in mating between the sexes.

Gaulin suspects that early humans were 'mildly polygynous' – some men had more than one wife, some had one wife, and others had none – and may have had a social structure somewhat similar to our closest living relatives, the chimpanzees. In the woodlands and savannas of Africa, chimps form communities with a dozen to a hundred or more members, each with its own territory. Inside that range the females often move around independently of the males, a single female travelling with only her children as company. She generally covers less area than the males, who travel in twos, threes or fours, patrolling the entire range. Although females are promiscuous and mate with any male, the high-ranking males in the troop attempt to monopolize females around the time of ovulation and thus father most of the offspring. If humans had a similar social structure at some point in our distant past, the sex difference in spatial ability may have arisen from it rather than a sexual division of labor.

No one can go back in time to check Gaulin's idea, but there is one test that could be done now. His theory predicts that because male chimpanzees cover much more ground than the females, chimpanzees should have a sex difference in spatial ability. Since researchers have recently developed ways to measure spatial ability in monkeys, somebody will probably look for such a sex difference soon.

Gaulin's suggestion that the 'natural' human sexual arrangement is for males to sometimes have more than one wife is, not surprisingly, controversial, and the evidence for it is inconclusive. Among humanity's closest relatives besides the chimpanzee, gibbons are monogamous and gorillas are polygynous – dominant males form harems of several females. In general, Gaulin notes, in those primate species that are monogamous, the males and females are very nearly the same size, while in the polygynous species males are larger than females. Over the past two million years, the ancestors of modern humans evolved so that the size difference between males and females got progressively smaller, which implies to Gaulin that our distant ancestors were polygynous and moving toward monogamy – but hadn't quite got there.

Human history includes both monogamous and polygynous cultures, but there have been few 'polyandrous' societies – those in which females have more than one mate. Gaulin points out that if you examine the 1,154 human societies described in the *Ethnographic Atlas*, you'll find that 980, or nearly eighty-five percent of them, openly practice polygyny (men having more than one mate) to one degree or another. It seems that humans can be relatively happy in either a monogamous or polygynous culture, regardless of what the 'natural' arrangement may be – another illustration that predispositions are not fixed, but can be shaped by environment.

Working from this evolutionary framework, researchers have offered a variety of explanations for almost any human sex difference you can imagine: males are more aggressive because they were responsible for defense of the group and/or because they contended with each other for dominance and women; females are more nurturing because they were primarily responsible for taking care of small children; males are interested in younger females because younger women were likely to be healthier and bear children more easily; females prefer males who are good providers because they could have given more help in making sure their children survived. It's an approach that some dismiss as 'evolutionary story telling' and that others defend as good science. Gaulin points out, for instance, that in the case of his voles, he was able to form a hypothesis about why a sex difference appeared and then test it rigorously by predicting what he would find in each of several species.

But the specific details about why this or that sex difference appeared really aren't that important. Few people besides scientists really care whether women developed their verbal advantage to help them interact with their children, to assist in remembering food sources or merely as a by-product of their superior fine motor skills. For our purposes, the important lesson is that the sex differences we see in humans today are leftovers from our past, predispositions that were important for survival fifty or a hundred thousand years ago.

And herein lies the irony at the heart of the study of sex differences. There is a reason, a purpose for the differences between men and women, those things that make life so interesting, but the reason lies in the distant past. Long ago it made sense that men and women should have different abilities and different psychological traits, but we inhabit a modern, industrialized world now. As Steve Gaulin says, 'Evolution prepared us for yesterday, but we have to live in today.' There's no need for men to have an easier time reading a blueprint than women, but there it is. There's no need for women to speak more fluently or have an easier time learning to read, but there it is. There's no need for men to be more aggressive, women to be more nurturing, men to be more attuned to hierarchies and dominance, women to see relationships more in terms of a web of friendships. In fact, it sometimes seems as if things would be a lot more comfortable if some of these differences didn't exist, but we're stuck with them.

The flip side is that these differences may well make us – the human race – stronger. As Janice Juraska points out, humanity has always relied on its ability to adapt to new environments and to learn how to handle new situations, so diversity has always been important to us. And one of the ways the human race has ensured diversity is to have two sexes. 'As a species we've really conquered this planet,' she says, 'and a lot of that is thanks to "male" traits. But we've come to a point where there's no more space to conquer, so a lot of the things that females do well – the ability to co-operate, the ability to live in cohesive groups – may become more important.'

In short, if one way of thinking can't find a solution, perhaps another can. And sex differences, even if all the other reasons for their existence are gone, still make us more diverse. Think of it as one more gift of evolution.

CHAPTER 11

Where Do We Go From Here?

In the house next door, two beautiful children are growing up. They both have big blue eyes, strawberry blonde hair and faces like the cherubs in those old Renaissance paintings. Mara Rose, the baby, had her second birthday a few months ago, and her brother, Grant, will be four a few months from now. They're often in our home because my wife has the unshakeable conviction that any neighborhood children are her children too. (A few weeks ago Mara told her mom that 'Amy Pool is the best neighbor in the whole world.')

Grant and Mara are being raised in what their father calls 'a non-traditional household'. Dad, a writer, works at home most days, while the kids watch Mom head to the office wearing a business suit and carrying a briefcase. Dad does much of the cooking, including fixing breakfast for the children, and most of the laundry. Mom is more likely than Dad to take off on a business trip for a week or two. And being sensitive, socially conscious parents, the two of them have worked hard to raise their children in a gender-neutral environment. Grant got kitchen toys and Mara Rose got cars and trucks. Neither got guns. There's no teaching that 'Girls can't grow up to be firemen' or 'Boys don't cry'. There's no pressure to take on stereotyped sex roles.

So why is it that these two children are growing up so differently – and in such sex-typical ways? Their father tells me that when he cooks Mara will put on an apron and 'help', while Grant 'sits at the table and waits for the food'. Mara loves to try on clothes and apply make-up. Grant cares nothing for clothes and would rather die than put on a little lipstick. Mara loves to play with the telephone, Grant likes little cars. The two of them have a toy kitchen with a large plastic oven and stove. Mara uses it to cook food. Grant turns the oven on its side, opens up its door and uses it as a cave or some other fantasy abode.

Just the other day the two were visiting in our basement with Amy, who entertained them with some crochet work she'd been doing. Mara was fascinated and wanted to watch how Amy's hands worked the crochet hook and see how the yarn was woven into its pattern. Grant was bored and, grabbing the ball of yarn, he took off around the basement at full tilt, unwinding the yarn as he went and leaving a trail that eventually circled the room and wrapped around a column.

When friends dropped off their three-month-old with Grant and Mara's parents for a few days until they resolved some child-care problems, Mara was captivated by the infant and wanted to feed it and change it. Grant hardly noticed. He was more interested in putting on his fireman's hat, getting in his little fire truck and extinguishing imaginary flames. 'If the house caught fire,' his dad told me, 'Grant would "rescue" the baby. Otherwise, he's not interested.'

Grant (remember, he's not yet four) has a crush on a little girl at his preschool, and he's already fantasizing about how he could win her heart. 'Daddy,' he said to his father one day, 'if dinosaurs got Rachel, I would save her.' And what would happen then? his father asked. 'She would cook dinner for me.' His dad chuckles and says he isn't sure where this particular daydream came from – especially the part about dinner since that's mostly daddy's job at his house – but one thing is clear: it's a male fantasy. No little girl would dream up something like that about a boy.

Having seen it firsthand, Grant and Mara's parents say they no longer have any doubts that males and females are born different, and from what I can tell their experience is typical. They belong to a generation of idealistic parents who thought they could get rid of sex differences by providing a non-sexist environment for their children and who have discovered that the children have other ideas.

Christina Williams, the Columbia University researcher who traced out sex differences in how rats learn their way in a maze, says she often finds her eighteen- to twenty-one-year-old students skeptical of innate differences between males and females, but when she talks to alumni groups, who have seen their children grow up, 'they never have any trouble believing the sex difference stuff. They think it's nice that science has confirmed what they've always known.' Williams tells of an old hippie friend who never wore dresses herself or worried much about appearance but now has a little girl who is fascinated with wearing pretty dresses and arranging her tresses. 'She spends hours in front of the mirror getting the barrette just right in her hair,' Williams says. The ex-hippie mother is not amused.

Sheri Berenbaum, the Chicago Medical School researcher who studies CAH girls, says she has plenty of friends who, like Williams' buddy, went to college in the sixties and were convinced that sex differences were all a product of socialization but who have changed their minds as they've watched their children grow up. These people are not controlling parents, she says. 'They pretty much let their children do what they want to do.' But even though they do very little to force their children into sex-typical patterns, the sex differences in play behavior, toy preference and temperament seem as great or nearly as great as they were a generation ago.

None of this should be surprising, of course, given what sex difference researchers have been discovering over the past decade. With the wealth of evidence now available, anyone who still insists that boys and girls are born with exactly the same predispositions and potentials is wearing ideological blinders.

Let's stop for a moment and review what's known about sex differences. Thanks to sex hormones in the womb and later, males and females differ in a number of predictable ways. Physically, men are larger and stronger than women, while women are healthier and live longer. Women have better manual dexterity and more accurate control of small muscle movements, while men have better control of large motions. Men have slightly better vision, particularly for moving objects, while women have more acute hearing, smell, taste and touch. Cognitively, women excel on verbal skills, certain types of memory, and perceptual speed (quickly picking out details from drawings or strings of letters or numbers), while men take the lead in spatial ability and, perhaps because of the spatial advantage, mathematical reasoning too.

In the first decade of life, girls are more interested in people, boys in things (a difference that persists into adulthood), girls mature faster, boys engage in more rough-and-tumble play, and the sexes choose different types of toys to play with. At play, young girls are more focused on what they're doing, spending more time at an activity and being more likely to finish, while boys are more easily distracted, get bored with an activity more quickly and often drop what they're doing before it's done. Later in childhood, boys prefer more complex, structured games, particularly those with clear winners and losers, while girls choose less complex, more co-operative and less competitive activities. In adulthood, men are more aggressive, women more nurturing, women more interested in how well they do a job and men more concerned with their performance relative to others. Ethically, males are more likely to base decisions on a set of rules for determining right and wrong, while females tend to consider the personal consequences for the people involved. In conversation, women seek to make connections, while men exchange information and establish their positions relative to others.

None of these sex differences is completely fixed by biology, of course. The most that biology can do is establish predispositions which interact with the environment to create a person. Some of those predispositions – such as the penchant of boys for rough-and-tumble play – are certainly very strong, but others – the greater male propensity for math, for instance – are weaker and can be greatly modified by socialization. Hormones may push males and females in different directions, but society can either exaggerate or dampen these differences, depending on how we teach our children and what we expect of ourselves.

The differences are relics of the past, echoes of evolutionary pressures that pushed men's and women's bodies and brains in somewhat different directions. They have no real purpose now, except perhaps in the few dying hunter-gatherer cultures left in the world, but we're stuck with them.

This growing understanding of the sexes certainly seems to have come at a good time. Over the past few decades our society has been changing rapidly, with the roles of men and women clearly metamorphosing – but into what? We sure won't be going back to that 1950s "Father Knows Best" world where daddy went off to work while mommy stayed home with the kids and baked pies, but past that, nobody knows what the world we're building will look like once the construction dust settles. Furthermore, there's little agreement on what that world *should* look like.

Should it be unisex, or should there be differences between men and women? If so, what types of differences? The research into sex differences may not answer these questions, but it does give us a better idea of what we have to work with.

I must confess to a very personal interest in all of this. Amy and I are expecting our first child shortly before this book is due to appear and, like new parents everywhere, we're full of questions about getting this new life started out on the right foot. What should we expect with a boy? With a girl? What toys should we buy? What parenting techniques should we use? Will it hurt to put our little girl in pink or little boy in blue? Should we give our daughter extra help so that she doesn't fall behind the boys in traditional male areas such as spatial ability or higher math? Should we pay extra attention to our son's language skills so that he's not at a disadvantage to the girls?

I hear such questions often from parents. Our neighbor across the street has a couple of girls, one seven and one four, and already she's worried that her second-grader might fall behind the boys in math. The girl does fine at math now, her mom says, but the little boys all seem to spend their time playing video games, something that has no appeal for her daughters. By the time they get to high school, my neighbor wonders, will the boys have an advantage in such things as geometry and trigonometry because of all the time they've spent on video games? And what can she do about it?

For advice I approached some of the sex difference researchers who are mothers themselves. How, I asked, had their knowledge of the sex difference research affected the way they raised their own children? Their answers blended science with mothering experience and common sense.

'I'm very conscious that my little boy might be quite active because he's a little boy, and it is very natural and healthy,' said Laura Allen, the discoverer of a number of sex differences in the brain and mother of a boy who was three at the time of our interview. 'I think that if I weren't aware that boys and girls do play differently, and that there's probably a biological basis for this, I might be a little less patient. I might be a little more critical when he takes things apart. Really, he's doing what's very normal and very healthy for him.'

And Allen, six months pregnant with a little girl, was already thinking about how she would raise her daughter: 'I think I will give my girl all these toys that a little boy has and make sure she's very mechanical because I'm not very mechanical and I don't know if it's just because I wasn't given little boys' toys. Some of these toys that little boys have probably help a lot [to develop spatial and mechanical ability].' But what would she do, I asked, if her daughter didn't like trucks and building blocks? 'If the little girl refuses to play with toys that are made for little boys, do you push her into it? I don't know. What happens if you force her to play with these toys? Will she rebel?' It's still too soon to know. Her daughter, Angela, born in the fall of 1992, is a little young to be playing much with toys that develop spatial ability, but Allen will make sure that she is given every chance that her brother was.

As a mother of seven, Iowa State University psychologist Camilla Benbow certainly has had more direct, personal experience with children than any other researcher I spoke with, and her philosophy is simple: give each of them the chance to develop to the best of his or her ability. 'Having had seven kids, I can say that when they come out they have their own personalities – they're all different,' she says.

'A good mother shapes her behavior to match each child. I don't treat my boys and my girls the same way, but I don't treat each of my boys exactly the same way either. I'm trying to meet their needs, I'm not trying to make them the same.' One of her sons, for instance, told her he'd like to study ballet. 'Fine.' Another wants to do karate. 'That's fine too.' She has given her boys dolls and her girls trucks, but she's more concerned with providing them with educational toys. Learning is taken very seriously in the Benbow household, and learning is something that both sexes do equally well.

In school, Benbow says, the same principle should govern: boys and girls should be treated as individuals, each encouraged to learn and grow according to his or her own abilities and interests. Benbow's studies of mathematically precocious kids have convinced her that, statistically speaking, boys make up a greater percentage of the true math geniuses than girls and that the difference is to a large extent biological – implying that in this one area at least, equal treatment for the sexes will probably not result in completely equal outcomes. But even here, she says, girls should still be encouraged just as much as boys to take math courses and aspire to careers in science or mathematics. If they're not, then some of the girls who have the natural ability and interest to pursue these careers may never discover their potential.

As an example she points to a lesser known side of her Study of Mathematically Precocious Youth. Besides identifying and testing bright students, the program holds intensive summer courses that teach the junior high school kids – both boys and girls – an entire year of highschool physics in three weeks, with the objective of boosting their interest in science and preparing them for college. 'If I didn't believe environment was important I wouldn't be in the area I am, running these summer programs,' she says. The summer camps, which the students attend when they're thirteen or fourteen, do appear to have long-term effects on school performance and eventual careers, Benbow says, and the camp kids are more likely to go into math and science fields than those mathematically precocious students who don't attend camp.

This is one opinion on which I found complete unanimity among sex difference researchers. There is so much overlap in abilities and interests between the sexes that it's ridiculous to assume a child can't do something because he or she is the wrong sex, and it is damaging to discourage the child from trying. Give your little girl trucks and building blocks so she has a chance to develop her spatial ability. If she doesn't like to play with them, what have you lost? Offer dolls to your little boy. If he has a nurturing side, it doesn't mean he's any less a boy. And, Lord knows, men in this liberated era – when husbands are learning to share in the taking care of children – need all the help they can get.

In school, no child should be pushed toward one subject or away from another because of sex. My wife remembers being the best math student in her school up to seventh grade (her fifth-grade teacher called her 'Little Miss Perfect' for always getting 100s on the exams), but eventually she stopped trying. The reason: although she got plenty of encouragement at home in other academic areas, her mother insisted that 'Girls in our family can't do math.' Now in graduate school, Amy is trying to make up for lost time. And it's not just individual females who are damaged by these attitudes, but society as a whole. In times past, when women were expected to marry and raise a family, nothing more, perhaps society could get away with discouraging girls from learning math, but in a world that depends increasingly on educated women in the work-force, such discrimination carries a heavy penalty.

These are the minimum prescriptions, equivalent to the doctor's primary rule, 'First, do no harm,' but it may not be enough to treat boys and girls just alike in school. One of the most important results to come out of the study of sex differences is the discovery that males and females often learn and think in different ways. In order to give both boys and girls the best possible opportunities to learn and to succeed, it may be necessary to find new teaching methods.

Diane McGuinness points out, for instance, that at preschool and early elementary ages, boys have a very different pattern of behavior than girls. Boys are much more active and wander around more, have shorter attention spans, are more easily distracted and are less likely to follow suggestions from an adult. Most elementary school classrooms,

McGuinness says, seem to be designed more for little girls than little boys, and many boys may end up being classified as behavior problems when they're really quite normal (for boys) and just need a different approach.

Several researchers have suggested that it may make sense to use different methods to teach boys and girls to read. Sandra Witelson, for instance, argues that since reading involves an interplay between visual-spatial skills (recognizing the shapes of letters) and language skills, and since male and female brains seem to arrange spatial and language processing in different ways, then such teaching methods as phonics or look-say may work better for one sex than the other. In one study, men and women were asked to mentally go through the alphabet and count those capital letters that had either a curve in their shapes, like O, or that were pronounced with an 'ee' sound, like T. When searching for a shape, men were faster and more accurate, but the women did better in finding the sound – a result that implies men have an advantage in processing the shapes of letters and women have an advantage in processing their sounds. These different strengths may make one reading method easier to learn for males and another easier for females, although no one has done a comparative study to pinpoint the best methods for each sex. With the correct approach, it might even be possible to reduce the large number of boys with reading problems – if those problems are partly due to teaching techniques that are poorly matched to how the boys learn best.

It may also be possible to come up with ways of teaching mathematics at the high school and college level that prevent women from losing so much ground. Since girls outperform boys in math through elementary school and into junior high, some researchers have suggested that the boys get an advantage when math courses begin to tackle subjects that have a spatial component, such as geometry, trigonometry or calculus. Boys may be more likely to use spatial reasoning, a more efficient technique in such subjects, while girls try to reason through the problems verbally, which doesn't work as well. The verbal approach may suffice for arithmetic and algebra, so female students learn to rely on it, but when they're faced with geometry and trigonometry they find it difficult to develop the spatial reasoning that male students have used all along. If so, then perhaps girls – and some boys – would benefit from an earlier emphasis on spatial reasoning.

In general, many researchers suspect that boys and girls have very different learning styles and respond to authority and teaching in different ways. Two Stanford University researchers, Robert Hess and Teresa McDevitt, studied how mothers taught their four-year-old children to do a simple block-sorting task and rated the mothers' teaching style on direct commands ('Put that block over there') versus inviting the children to make their own contributions ('Where do you think this should go?'). They then gave the children achievement tests, and followed up with two later rounds of tests, once at ages five or six and again at twelve. Consistently, the children whose mothers involved them in the learning process by inviting their opinions scored higher than the children with controlling mothers, but a curious pattern emerged. The girls were much more influenced by their mothers' teaching styles than the boys, and the influence decreased over time for the boys but not the girls. Indeed, by the time the boys reached twelve years old, they were basically unaffected by whether their mothers were the types to use direct commands or a more collaborative approach. The moral: when mothers (or fathers or anyone else) teach girls, they should be particularly careful to involve them directly in the learning instead of simply telling them what to do. This doesn't seem so important with boys, however, perhaps because they're less sensitive to others' behavior.

One study of mathematics teaching techniques in grade school found that girls performed more poorly in those classrooms where the teacher emphasized competition between students, while boys did slightly better in competitive settings. On the other hand, girls' math performance improved when the teacher used co-operative methods in which students worked together to get answers, while such co-operative teaching techniques neither helped nor hurt boys' performance. Another study reported that teachers who encouraged autonomy in their students inspired girls to take on more challenging work than teachers who were very controlling, but it saw no difference between the two teaching methods for boys.

Indeed, a number of researchers have suggested that males' greater autonomy in learning may play a large role in why boys do better on standardized math tests while girls get better grades in class. According to this idea, boys are inherently more rebellious and independent in the classroom while girls are more concerned with good behavior, so if a teacher tells the students to work a problem one way, the girls will learn to do it that particular way but the boys will look around for other ways to get the answer. This leads to girls learning things more by rote and to boys creating their own independent ways of solving things. On classroom tests, where you're expected to solve problems just like those already presented, girls would have an advantage, but on standardized tests and in other situations where material is new or presented in a different way, boys would be better equipped to handle it.

If this is correct – and evidence from a number of directions suggests that it is – teachers should be particularly diligent in infusing girls with a sense of independence in learning. Boys are going to march to their own drummers one way or the other, but girls are more willing to behave as others wish.

Ultimately, recognizing that boys and girls learn in different ways and understanding what those different ways are will lead to improved teaching techniques for everyone, male and female. Some methods will work better for boys, on average, and some will work better for girls, but knowing what we do about the overlap between the sexes, it's certain that some boys will learn better with 'female' teaching methods and some girls with 'male' techniques. The trick will likely be to learn what methods work for each sex and then use all of them in some sort of mixture.

Everyone agrees: boys and girls should be given the same opportunities to learn and develop skills as they are growing up, and later in life men and women should be given the same economic, political and social opportunities. Indeed, it's hard to find anybody in this country who doesn't believe in this 'equality of opportunity'. But is it necessary that men and women have the same jobs, make the same money, hold the same number of political offices, and change the same number of diapers? When you start talking about this 'equality of outcome', that's where disagreement sets in.

Many feminists in the 1960s and 1970s (and many still today) assumed that equality of opportunity and equality of outcome were basically synonymous. Since, as they believed, there were no natural differences between the sexes except for some unimportant physical ones, then if males and females live in a non-sexist society, equality of outcome would follow automatically. Conversely, if there were any statistical inequalities between men and women in jobs, educational achievement, economic power, political office-holding, child-raising or other areas, they must be caused by sexism and must therefore indicate that males and females were not being treated equally by society. To measure progress toward sexual equality, all one had to do was measure the numbers of men and women in various jobs, political offices and so on – if those numbers weren't equal, sexual discrimination in one form or another must still exist.

Some feminists still hold this position, but more and more are concluding that equality of opportunity is not going to automatically produce identical outcomes between the sexes. Sheri Berenbaum is a good example. A feminist, she has done as much as anyone to demonstrate the power of hormones in the womb. Her studies of CAH girls are probably the most convincing evidence to date that hormones during development create different dispositions in males and females, including toy preferences and spatial ability. So over lunch, I asked her whether equality of outcome between the sexes should be a goal of society.

She has no objection to men and women being unequally represented in different professions, she told me, as long as those professions are equally respected and valued. In principle, it's OK if doctors tend to be men and nurses tend to be women, but 'as long as people value physicians more than nurses, we have a problem.' This is the reason, she believes, that many feminists call for equal numbers of men and women in different professions. 'It's easier to argue for equal representation than for equal valuation.' As long as some professions have higher prestige, more respect and higher pay than others, Berenbaum believes that we should try to even out the sex differences in those jobs as much as possible.

In particular, we spoke of Camilla Benbow's findings that boys make up a disproportionate percentage of the top junior high students in mathematics – and thus a large proportion of science and engineering students later on. Berenbaum accepted the possibility that the difference might be partially due to biology, but maintained that since science and engineering jobs are high-profile, high-prestige occupations, it's desirable to try to equalize the numbers of men and women in them. This might entail some type of educational intervention to improve girls' spatial ability or extra efforts in teaching high school and college mathematics to females. And it would probably demand programs to attract women to science and engineering careers to stop the leakage of women from the science 'pipeline'. Studies have shown that throughout high school and college, females with an initial interest in science are more likely than males to turn to other subjects. 'There are probably differential selection pressures on men and women,' Berenbaum said. Whatever the reasons for the imbalance, a more uniform sexual mix in these occupations would be a worthy social goal, she said.

Benbow disagrees. She has no problem with numerical inequalities between the sexes – even if more men hold the higher prestige, better paying jobs – as long as the disparities arise naturally from differences in abilities and interests and are not the product of discrimination. If fewer women than men want to be engineers, or have the spatial ability necessary to be engineers, why try to even out the numbers artificially? It's a *laissez-faire* attitude – let individuals have the freedom to make their own choices and don't worry about whether this group is making exactly the same choices as that one. Although Benbow is solidly behind equal opportunities for males and females (I'd hate to be a teacher who told one of her daughters that girls can't do math), she doesn't expect equality of opportunity to result in equality of outcome. Men and women have different interests, different values, different abilities and different goals, and you should assume those differences will result in men and women making different decisions.

Berenbaum and Benbow are representative of two views of sexual equality, a group-based view and an individual-based view. Do you measure equality in terms of group outcomes, asking that groups be statistically alike in jobs or pay or 'value', or do you define equality as each person having equal opportunity? There's little common ground between these points of view, and the choice of one or the other will point a society in very different directions.

But even supposing we can agree on what sexual equality is, it's going to be tough to know when we've reached that perfectly sexually egalitarian world. Of course, if your definition of a non-sexist world is one in which there is complete equality of outcome, that's easy to measure: do we have fifty female senators? Is every other doctor, lawyer and college math professor a woman? Do Hollywood's leading ladies get paid as much per film as its leading men? But traditionally, the United States has never demanded equal outcomes, only equal opportunity, and that's much harder to gauge.

Some researchers, such as Benbow and Doreen Kimura, suspect that if people simply follow their natural inclinations, then sizable sex disparities in the workplace and elsewhere will emerge – and, indeed, that there is much more equality of opportunity today than one might guess simply by looking at the numbers of men and women in various occupations. In an article in the September 1992 *Scientific American* Kimura concluded,

'The finding of consistent and, in some cases, quite substantial sex differences suggests that men and women may have different occupational interests and capabilities, independent of societal influences. I would not expect, for example, that men and women would necessarily be equally represented in activities or professions that emphasize spatial or math skills, such as engineering or physics. But I might expect more women in medical diagnostic fields where perceptual skills are more important.'

Other researchers, however, argue that today's sex differences in occupations are mostly the result of women being pushed in certain directions, covertly by socialization and overtly by sexual discrimination. UCLA's Melissa Hines, for instance, sharply challenged Kimura's conclusion in a letter to *Scientific American*, calling it 'misleading and potentially damaging'. Sex differences in abilities, she wrote, are not nearly large enough to explain the differences in occupations. 'Researchers studying sex segregation in occupations have concluded that the major determinants are economic and political, not hormonal,' and '[i]f women continue to be misinformed about their chances of succeeding as engineers and scientists, the sex ratios in those occupations are unlikely to change.'

Not so, Kimura responded. 'The common inference that women are kept out of the sciences by systemic or deliberate discrimination is not based on evidence. One might as well argue that men are kept out of nursing careers by discrimination. Instead the process appears to be largely self-selection.'

And as she does so often in her own research, Kimura has put her finger on the key issue: self-selection. Certainly there are relatively few jobs for which one sex or the other is clearly better suited in terms of natural abilities. Men are better equipped for occupations that demand physical strength – farming, construction work, professional wrestling – and they probably have an edge in careers that require spatial skills, such as architecture or engineering, and perhaps also professions that need high-level mathematical ability, such as mathematics or theoretical physics. Women, as Kimura pointed out, should have an advantage in jobs that require perceptual speed, such as medical diagnostics, and their verbal skills may give them an edge in such professions as writer, editor or newscaster. But for most occupations there seems to be no reason – as far as physical or cognitive abilities go – that one sex or the other should predominate. Why, for instance, aren't there more women accountants?

If the reason is sexual discrimination, then we need to get rid of it. But suppose the difference in the numbers of men and women in various jobs is due simply to the fact that – for whatever reasons – men and women make different choices about what they want to do with their lives. If that's the case then many would argue that we, as a society, should not try to change it. After all, do we really want a government program to push more women to become doctors and more men to be nurses if members of both sexes are just doing what they want to do? So before we start making any policy decisions we need to know: how much of the sex difference in occupations is due to the sexes having different goals and preferences and thus choosing, on average, different types of jobs?

Quite a lot of it, says Camilla Benbow. Among her mathematically precocious youth, the boys and girls are quite different in what they want to do with their lives. The boys are 'very theory-oriented' and tend to be single-mindedly interested in math and science, but the girls are just as likely to be drawn to social and artistic pursuits as they are to math or science. 'They're more balanced,' Benbow says – they like everything. So when the students get to college, most of the boys choose majors like mathematics, physics or engineering, while the mathematically talented girls spread out over a number of areas.

A similar difference in interests pushes men and women to pursue careers in different scientific and technical areas. The more 'people content' there is in a field, the more women are drawn to it. Women outnumber men among those getting degrees in psychology, social work, education and health sciences. Men predominate in engineering and architecture, physics, chemistry, math, and theoretical science of almost any sort – in general, those subjects that deal with things or abstractions, not people.

Even within a given subject, the sexes often choose to go into different specialties. When visiting Benbow at Iowa State, for instance, I noticed that most of the people in the psychology department were men. Since I knew that about sixty percent of people earning psychology Ph.D.s today are female, I asked Benbow why there weren't more women on the faculty. Among Ph.D. psychologists, she explained, women are more likely to go into clinical psychology – where they can help people – while more men choose to go into psychological research. 'Women are making very different choices than men, and it relates to differences in preferences and values,' she said. 'Women are drawn to people-oriented fields, men toward objects.'

And it's not just in scientific and technical fields. Many of the male–female divisions we see in the workplace follow along this same person/object divide. The more contact with people and the more opportunity to help individuals, the more likely women are to predominate in an occupation. Nursing and teaching are two good examples. On the other hand, if a job entails making or fixing something – the building trades, for instance, or car or appliance repair – then you're likely to find a lot more men doing it.

There's actually not much debate on the existence of this self-selection phenomenon, since survey after survey shows that from the time they're quite young, males and females want to do different things when they grow up. But feminists insist that even if boys and girls do differ in career goals, those differences are exaggerated by discrimination in the work world. Many women scientists, for instance, feel that they're treated differently – not taken as seriously, held to a different standard – than their male colleagues. And furthermore, some feminists say, the sex difference in career choice is created, or at least greatly magnified, by the subtle and not-so-subtle messages that children get from the time they're born about the proper roles of men and women. Many women scientists, for instance, feel that they're treated differently – not taken as seriously, held to a different standard – than their male colleagues. And furthermore, some feminists say, the sex difference in career choice is created, or at least greatly magnified, by the subtle and not-so-subtle messages that children get from the time they're born about the proper roles of men and women.

In response, some researchers point out that the sex difference in interests does not have to be created – it exists almost from the time a child is born. In the first year of life, boys and girls already respond differently to people and objects – girls smile and vocalize at pictures of people but not at things, whereas boys are just as happy to look at inanimate objects or blinking lights as they are to look at people's faces. By two, the difference is well established. In one experiment, for instance, a researcher asked two- to four-year-olds to illustrate a story. About half of the girls put people in their pictures, but less than one in five of the boys did. They preferred cars, trucks and fire engines.

Debates like this can go round and round, like a dog chasing its tail, until everyone is dizzy but nobody has gotten anywhere. What type of sexual equality should we aspire to? How close could we get to equality of outcome if we truly had equality of opportunity? How will we know when we've vanquished sexual discrimination? What do men and women really want?

The only way we're ever likely to find out is to just do it – create a new society from scratch that throws out all the old stereotypes about sex roles and teaches children that there are no differences in the things men and women should do. Release women from their child-raising responsibilities and allow them to contribute to society in ways no different from men. Make sure that males and females have exactly the same legal, economic and social power. Then we could see once and for all just how close we can come to the non-sexist society envisioned by many feminists.

Sound like an impossible dream? It's not really. Over the past seventy years in a land halfway around the world, just such a grand sociological experiment has been unfolding. Has this new egalitarian society succeeded in eradicating sex differences in social roles? Do its people live in a unisex utopia, free of inequalities between the sexes? Yes and no, as we shall see. Anthropologists have studied this custom-made culture intensely, and it offers a number of lessons about the type of sexual equality our own society can hope to achieve.

In Israel, the kibbutz has now been home to three generations of men and women. Founded in the 1920s by radical young European Jews, many of them Marxists, the kibbutz movement was intended to create a new type of society – a communistic one where the property is owned jointly and the fruits of labor are divided evenly, a democratic one where every adult has an equal voice in its direction, and an egalitarian one where males and females are equal not only in the eyes of the law but in each other's eyes as well. Today some one hundred thousand Israelis live in hundreds of these kibbutzim, or collective rural settlements, scattered across the country, following a way of life much different from their fellow countrymen.

From the beginning, the old ways were banished from the kibbutz. Capitalism is gone, replaced by a system of sharing in which everyone strives – in theory, and to a large extent in practice – for the good of the whole. Much of religion is gone, particularly the rituals that treated men and women differently. And much of the ideal of the family is gone, with children now seen as the responsibility of the community as a whole.

Indeed, the rethinking of parenting and housekeeping was one of the linchpins of the kibbutz. In his book *Gender and Culture: Kibbutz Women Revisited*, anthropologist Melford Spiro describes the ideology of the early kibbutz movement as dedicated to a view of sexual equality that accepted no differences between men and women except for the (unfortunately unalterable) fact that women had babies. Equality of opportunity was not for these idealists – they would be content with nothing less than men and women participating equally in all facets of kibbutz life, from plowing the fields and raising the children to holding political office and making decisions about the direction of the kibbutz. Spiro calls this the 'identity' version of sexual equality. In the view of the early kibbutzniks, the main barrier to this type of equality was the fact that women had the major responsibility for taking care of children. So one goal of the kibbutz movement was to liberate women from their traditional family duties – raising children, cleaning house and cooking meals – so that they could join men as equal partners, working alongside them in the fields, which was seen as the 'real work' of the kibbutz.

From the start of the kibbutz movement, children have been raised in dormitories. Starting at two to six weeks of age they live in these dorms where nurses and teachers, not their parents, are responsible for their care. Today there is a small percentage of the kibbutzim in which children do sleep in their parents' apartments, but even in those the child-care staff take care of children during the day.

Furthermore, neither husband nor wife need worry about who does the cooking, washing of clothes or sewing. Everyone eats in dining halls, where each member of the kibbutz takes turns cooking and serving the food. Clothes are washed, ironed and mended in communal shops.

The structure of the kibbutz was designed to ensure economic, social and political equality. No woman is dependent upon a man for support (or vice versa). Everyone who lives and works in the kibbutz receives the same rewards, independent of the type of job. Marriage, although permitted, has nothing to do with social status. A single mother of six is treated the same as a married woman or as a married or single man. Every adult kibbutznik has the same say in the kibbutz's affairs, and since the business of the kibbutz is overseen by a large number of groups whose personnel include up to half the members of the kibbutz, political power is widely dispersed among both men and women.

The first generation of Israelis to populate the kibbutzim were idealistic people, dedicated to political, economic and sexual equality and committed to making the new system work. And in many ways they have. They have gotten rid of private property and the unequal valuation of different types of work, they have created a non-religious society, they have maintained a rural, agricultural lifestyle, and they live in an exceptionally democratic polity. But, say Lionel Tiger and Joseph Shepher in their book *Women in the Kibbutz*, these efforts have fallen short in one obvious way.

Tiger, a Canadian anthropologist, and Shepher, an Israeli sociologist and long-time member of the kibbutz movement, undertook a massive, multi-year analysis of the role of women in the kibbutz, and what they discovered startled them. Despite the success of the kibbutz in other areas, and despite the fact that the young adults in the kibbutz today grew up there, being indoctrinated with the ideology of the movement, the kibbutz's goal of creating a society where men and women work at the same jobs, hold the same power and have the same attitudes toward family and career has failed.

Today, most occupations in the kibbutz are clearly either 'men's work' or 'women's work'. Men outnumber women by nearly seven to one in agriculture, and there are almost no women in trades such as carpentry, plumbing and electrical work. The male/female ratio is more than two to one in industry and management positions and it is three to one among the few people involved full-time in public service. Conversely, there are six times as many women involved in service and consumption jobs, and nine times as many in teaching. And inside these categories the sex distinctions are even sharper. In teaching, for instance, there are virtually no men who work with preschoolers but about forty percent of the high school staff is male. Men do essentially no care of infants. And of the twenty-five percent of the management workers who are women, most are secretaries, typists and accountants.

Although males and females take the same amount of schooling on average, women are much more likely to get degrees for teaching or nursing, while men aim for careers in agriculture, engineering and management. Women are less active in politics, and the higher the office the smaller the percentage of women occupying it.

Among married couples, Tiger and Shepher report an equality that Western societies could learn from. The husband and wife have equal say in decisions that affect the family unit, and the housework – cleaning, repair work, doing a little cooking in the kitchenette – is shared more equally than in many Western countries. But it is not completely equal. On average, Tiger and Shepher estimated, the wife does twice as much work around the small apartment as the husband – eight hours, versus three or four. (Husbands, however, work a little longer on average for the kibbutz, so it all evens out in terms of total hours). The women do most of the cleaning, while the men do such things as electrical or plumbing repairs, heavy work in the garden and taking out the garbage.

Although the kibbutz began with the idea of supplanting the traditional family with the commune, over the years the family has evolved to become the most important social unit in the kibbutz, Tiger and Shepher report. Almost all members of the kibbutzim get married now. And although the children are still raised communally, mothers have pushed for greater and greater involvement in their upbringing. Upon the urging of women in the kibbutz, a special period called 'the hour of love' was set aside for mothers to take their children from the dorms to walk and talk with them. And in those ninety percent of the kibbutzim where the children sleep in dorms, Tiger and Shepher found the women urging that the children be allowed to spend nights in parents' apartments.

In short, from an ideologically unisex beginning, the kibbutz has developed a sexual division of labor almost identical to what is found in most societies. Furthermore, the family unit – husband, wife and children – came to take on more and more importance as time went on.

Do the members of the kibbutz – especially the women – believe that their efforts to achieve sexual equality have failed? Not at all, reports Spiro. Instead, they have changed their ideas about what constitutes equality. In place of the original 'identity' vision of equality, which insisted that men and women should do exactly the same things, the kibbutzniks now subscribe to an 'equivalence' view, where equality is assumed to exist if men and women have equal opportunities and are perceived to contribute equally, even if they make different choices and play different roles. The second- and third-generation members of the kibbutz no longer expect men and women to perform the same jobs, or think that role identity is necessary for the sexes to be equal.

One clear, although rather frivolous, sign of the shift in ideologies is the change in women's attitude toward appearance. Among the first-generation kibbutzniks, women shunned anything that might set them apart from the men, disdaining make-up and jewelry and wearing men's baggy pants instead of skirts and dresses. Now kibbutz women wear stylish clothes, jewelry, make-up and perfume no less than women outside the kibbutz.

But despite the reappearance of sex differences in jobs, political activity, family matters and even appearance, both male and female kibbutz members believe that there is true equality between the sexes,

Spiro reports. Economically, every member of the kibbutz still gets paid the same, no matter what job he or she performs. Politically, everyone has an equal say, and although more men than women are in positions of authority, that is because more men than women are interested in serving in these unpaid positions that demand extra hours beyond a person's regular job. No one sees women as having less political power than men. And socially, there is no difference in status or prestige between men and women. Although everyone acknowledges that men and women are likely to end up in different careers, very much divided along the traditional male/female pattern, members of the kibbutz believe that jobs are equally open to all, regardless of sex, and that neither sex is more or less valuable to the community.

Two lessons emerge from the kibbutz experience. First, it is exceptionally difficult to form a society that has sexual equality in its identity version, with men and women doing exactly the same things. The ideological, highly committed kibbutzniks succeeded in replacing capitalism with communism, they succeeded in transforming an urban, paternalistic society into a rural, sexually democratic one, but they failed in abolishing the sexual division of labor and the mother-child bond.

But second, it doesn't really matter. The kibbutzniks decided that sexual identity wasn't necessary for sexual equality after all. In their world, men and women are equal in the ways that matter, even if they don't do exactly the same things.

Can we achieve the same type of equality in the United States and other Western countries? We certainly will never have the same built-in economic equality, given our capitalistic economies. With the kibbutz method of rewarding everyone equally no matter what jobs they do, there can never be a pay differential between the sexes, even if men cluster in some jobs and women in others.

On the other hand, the sexual division of labor in the kibbutz seems to be exaggerated by the special conditions there, and is probably much higher than Western countries can hope to achieve. Each kibbutz is basically a small agricultural commune with between a hundred and a thousand people. Some of them have manufacturing, but the main business of each is farming, which directly involves about a fifth of the workers on each kibbutz. Education and child care takes another fifth of the work-force, and the service-consumption sector (which entails such things as preparing food and taking care of clothes) demands about a quarter of the workers. This doesn't create a whole lot of career choices.

Men end up doing most of the agricultural work primarily because of their greater physical strength, not because they find field work more inviting and rewarding than women do. Indeed, Spiro reports that a majority of both sexes on the kibbutz are not happy with the range of jobs open to them – the kibbutzniks are generally well-educated but they seldom have much chance to put their education to work. Given their limited options, the men and women of the kibbutz do tend to segregate themselves into sex-stereotypic roles, but if they had the broad range of jobs that are available in an industrialized economy, they would probably spread out more.

Perhaps the largest difference between the kibbutz and Western societies, in terms of the practical effects for women, is the way the kibbutz handles child care. Two to six weeks after a woman gives birth, she's back to work and her baby is in a dorm, whether mom likes it or not. This enforces a symmetry between men and women that is impossible to achieve in a society in which the parents care for their children. Even though many of the women on the kibbutz are doing just what women in other countries primarily do – taking care of children, washing and mending clothes, preparing food – on the kibbutz it's a job, just like the jobs that the men and the other women do. There can be no division between the women who stay home to take care of their children and the men and women who hold jobs.

This solution sounds awful to me – I hate the idea of having so little contact with my child and so little say in how he or she is raised – but I do understand the problem that the founders of the kibbutz movement were trying to solve. The more I spoke with women for this book and the more my wife and I talked about the issue, the more convinced I became that there is a problem here. In fact, I've come to believe that the biggest challenge to sexual equality – in any sense – is the fact that women are the ones who have the children, and that women inevitably seem to be the ones who feel the most responsibility for taking care of them, especially in the early years. This puts a constraint on women that men just don't have.

One of the most poignant interviews I had was with Sandra Witelson, the McMaster University psychologist who studies the relationship between brain structure and function in men and women. Did she have any children? I asked. Yes, one. 'If I'd had a second and third child I'd probably never have gotten tenure,' she said. 'You can't get tenure by working nine to five because the men don't.

'I look around and see how many of my male colleagues have three or four children. I would love to have a "wife". I see this in my staff. When a child is sick, it's always the wife who stays home. And I don't think it's just because of society. I think they want to be home with the children.'

My own wife is wrestling with the same thing. She's just as ambitious as I am. She too gets a great deal of pleasure and satisfaction from work, and she too would like to feel she made her mark on the world with a career that really accomplished something. On the other hand, she really wants children, as I do. And once we have them, we both know that she'll have a much harder time being away from them than I will. We also both know she's the one best suited for raising the young ones – she has a knack with small children, and a love for them, that I can admire but never hope to match. (On the other hand, I do pretty well with kids past two or three, especially boys, and I'm sure I'll take on much of the responsibility for them.) So Amy is wondering how she can combine raising children with having a career, especially since she'll be competing with men whose wives take care of their kids for them. 'It's not fair,' she's told me many times. What can I say?

If we're going to have sexual equality approaching that of the kibbutz, my guess is that two things will have to happen. Since the United States isn't likely to set up a universal child-care system that sucks in children at two to six weeks of age and doesn't spit them out again until they're teenagers (although a lot of parents might support such a system if it took the kids once they hit the teenage years), we need to re-establish child-raising as a respected occupation. The kibbutz recognizes it as being just as essential to the community's well-being as farming or management careers, but the US is burdened with the 'just a housewife' syndrome. This attitude is a relic from a male-centered era (unfortunately and ironically exaggerated by the feminist movement) that valued only traditional male activities and relegated 'women's work' to a secondary tier. But if a person decides to take time off from a career – or even completely forgo one – to raise children, this shouldn't be seen as dropping out of the real work of the community. It's simply another job.

Second, we need to find some way that women (and men too) can integrate child care with career. In our hunter-gatherer days this was no problem – the women simply carried the children with them as they gathered food, contributing just as much to the family sustenance as the men did. But now our workplaces are set up on the assumption that a person will either have a job or take care of children, and that these two activities are and should be completely separate. This may have worked fine a couple of generations ago when the work-force was mostly men and single women (or at least women who had no children at home), but it's a problem now. What does a woman (or a man) do who wants to devote a lot of effort to being a parent but also keep up a career?

I do see some flexible solutions appearing. Employers are starting to allow new mothers longer leaves and part-time hours after they return to work. Some companies offer on-site day care so that their female employees can visit and nurse their children throughout the day. And individuals are creating their own arrangements. One friend of mine, for instance, works at home as a freelance writer so that she can spend more time with her young son.

Such solutions will become widespread only if men and women are accepted as equal but different. Demanding that women be like men if they are to pursue a career has left women like Sandra Witelson thinking they must choose between having successful careers and limiting their families, or having larger families and sharply limiting their career options. This will probably always be a larger problem for women than for men, since as sociologist Alice Rossi points out, the mother–child bond seems to be the basic family bond. Fathers can become just as attached to their children and devote just as much time to them, but especially with children under a year old, the mother appears naturally to be the one to take the primary responsibility. 'If a woman really wants a long-term career,' Witelson says, 'it is important to be aware of what it means to be a woman. It doesn't do a woman any good to tell her she's just like a man so she should go out and do it.'

I tell my wife that she is blessed to have so many strong desires and abilities. After all, some people have nothing they really enjoy doing or nothing they're really good at, while she loves to work, she loves to be with children, and she's very good at both. But we both know that it's as much a curse as it is a blessing. Sometimes, she says, she wishes she could be happy just raising a family in a nice small town somewhere. Other times she'll tell me she wishes she'd been born a man so she could follow her career goals at full speed with no questions, no regrets. This is not a dilemma that many men have.

My crystal ball is as cloudy as the next guy's, but I suspect that we're working our way toward a society that has a fair amount of sexual equality in the equivalence sense. I think the idea of creating sexual equality in the identity sense is a pipe dream – if a small group of ideologues building a society from scratch in the Middle-Eastern desert couldn't do it, how can we? Men and women aren't the same, and they don't want to be the same. But I detect a strong and growing commitment in our society toward making sure that the sexes have equal chances to 'Be all that you can be', as the US Army's catchy slogan puts it.

The lesson of sex difference research is that men and women are different, but still equal. A truly equal society will not be one in which women have turned out to be just like men, but rather one in which women have been freed to make their own contributions, to reach their own potential. With two sexes working together, complementing each other, it can't help but be a better, stronger world.

NOTES

INTRODUCTION – Hormones and Heroines

page
1 **the spotted hyena.** For details on the spotted hyena, see, for instance, David Macdonald (ed.), *The Encyclopedia of Mammals*, Facts on File Publications, New York, 1984, pp. 154–7.

one South African researcher... Norman N. Deane, 'The Spotted Hyena', *Lammergeyer*, Vol. 2 (1962), pp. 26–44.

2 **trouble telling the sexes apart.** For details, see Laurence G. Frank, Stephen E. Glickman and Irene Powch, 'Sexual Dimorphism in the Spotted Hyena', *Journal of Zoology, London*, Vol. 221 (1990), pp. 308–13. This paper provided some of the first methods for reliably telling male and female hyenas apart without a physical examination. They noted, for instance, that the tip of the penis has a slightly different shape in males and females. Frank estimates that 'experienced zoologists had spent over fifty observer-years watching hyenas in the wild' without spotting this relatively subtle difference.

an unusually high percentage of pups are born dead... Laurence G. Frank and Stephen E. Glickman, 'Giving Birth Through a Penile Clitoris: Parturition and Dystocia in the Spotted Hyena', *Journal of Zoology* (in press).

3 **the most aggressive infant mammals known to science.** Glickman and Frank note that they have not yet proved a direct cause–effect relationship between the testosterone in the womb and the aggressiveness of the newborn hyenas, although it is a plausible hypothesis. They plan to test the hypothesis with experiments in which they block the effects of the male hormones and see how this affects the hyenas' aggressiveness.

newborn spotted hyenas may face a fight to the death... Laurence G. Frank, Stephen E. Glickman and Paul Licht, 'Fatal Sibling Aggression, Precocial Development and Androgens in Neonatal Spotted Hyenas', *Science*, Vol. 252 (1991), pp. 702–4.

4 **no other mammal drops out of the womb ready to fight and bite.** Strangely enough, domestic piglets seem to be the closest thing to spotted hyenas in rivalry among newborn mammals. There are often more piglets than teats on the mother sow, and the piglets compete for places at Mom's Diner. A piglet's canine teeth are in place at birth and it uses these weapons to defend its rights to a particular teat. See David Fraser and B. K. Thompson, 'Armed Sibling Rivalry Among Suckling Piglets', *Behavioral Ecology and Sociobiology*, Vol. 29 (1991), pp. 9–15.

7 **'hidden agendas.'** Anne Fausto-Sterling, *Myths of Gender: Biological Theories About Men and Women*, Basic Books, New York, 1985, p. 9.

CHAPTER 1 – Different but Equal
page
14 **a provocative and troubling scientific study.** Camilla Persson Benbow and Julian C. Stanley, 'Sex Differences in Mathematical Ability: Fact or Artifact?', *Science*, Vol. 210 (1980), pp. 1262–4.

15 **a simple explanation for this difference.** This hypothesis was suggested in, for example, Elizabeth Fennema and Julia Sherman, 'Sex-Related Differences in Mathematics Achievement, Spatial Visualization and Affective Factors', *American Educational Research Journal*, Vol. 14 (1977), p. 51.

18 **'If your mother hates math...'** Sheila Tobias, author of *Overcoming Math Anxiety*, quoted in 'Do Males Have a Math Gene?' *Newsweek*, 15 December 1980, p. 73.

'darned shaky ground...' Elizabeth Fennema, education researcher at the University of Wisconsin in Madison, quoted in Gina Bari Kolata, 'Math and Sex: Are Girls Born with Less Ability?', *Science*, Vol. 210 (1980), pp. 1234–5.

'what we have here...' Elaine Newman, biology professor at Concordia University in Toronto, quoted in 'Boys Beat Girls in Controversial Study', *Toronto Globe and Mail*, 3 September 1981.

'It is a political issue.' As quoted in Philip J. Hilts, 'At Mathematical Thinking, Boys Outperform Girls', *The Washington Post*, 5 December 1980. pp. Al&A10.

'It is virtually impossible to undo the harm...' Alice T. Schafer and Mary W. Gray, 'Sex and Mathematics', *Science*, Vol. 211 (1981), p. 231.

19 **testing every subtle environmental influence.** For an exhaustive review, see Camilla Benbow, 'Sex Differences in Mathematical Reasoning Ability in Intellectually Talented Preadolescents: Their Nature, Effects, and Possible Causes', *Behavioral and Brain Sciences*, Vol. 11, No. 2 (1988), pp. 169–83.

'After fifteen years looking...' Anne Moir and David Jessel, 'Sex and Cerebellum', *The Washington Post*, 5 May 1991, p. K3.

20 **A likely explanation for boys gaining on the SAT verbal section.** Diane F. Halpern, *Sex Differences in Cognitive Abilities*, Lawrence Erlbaum Associates, Publishers, Hillsdale, New Jersey, 1992, p. 92.

A study done by Janet Hyde, Elizabeth Fennema and Susan Lamon. Janet S. Hyde, Elizabeth Fennema and Susan J. Lamon, 'Gender Differences in Mathematics Performance: A Meta-Analysis', *Psychological Bulletin*, Vol. 107 (1990), pp. 139–55.

This pattern of differences. Halpern, *op. cit.*, pp. 221–2.

21 **Teen Talk Barbie.** Actually Barbie was politically incorrect even before she could talk and complain about math class. That exaggerated hour-glass figure has been blamed for girls having unrealistic expectations about what they're supposed to look like, so in 1991 a competing doll was introduced, called 'Happy to Be Me'. Its major selling point was that at 36–27–38 it had a

figure much closer to what little girls can expect when they grow up. See, for instance, Cyndee Miller, 'Flat Feet and Big Hips – Now That's One Happy Doll', *Marketing News*, 30 September 1991, p. 2.

22 **This much of Binet's work is well known.** See, for example, Theta H. Wolf, *Alfred Binet*, The University of Chicago Press, Chicago, 1973, pp. 139–89.

McGuinness tells this Binet story. Diane McGuinness, *When Children Don't Learn: Understanding the Biology and Psychology of Learning Disabilities*, Basic Books, New York, 1985, p. 19.

23 **a study of language development.** Dianne Marie Daugherty Horgan, 'Language Development: A Cross-Methodological Study', Unpublished dissertation, University of Michigan, Ann Arbor, 1975.

Other researchers have found... Several of these studies are reviewed in David W. Shucard, Janet L. Shucard and David G. Thomas, 'Sex Differences in Electrophysiological Activity in Infancy: Possible Implications for Language Development' in Susan U. Philips, Susan Steele and Christine Tanz (eds), *Language, Gender and Sex in Comparative Perspective*, Cambridge University Press, Cambridge, England, 1987.

girls have translated their early language advantage... David J. Martin and H. D. Hoover, 'Sex Differences in Educational Achievement: A Longitudinal Study', *Journal of Early Adolescence*, Vol. 7 (1987), pp. 65–83.

A recent cross-cultural study. Virginia A. Mann, Sumiko Sasanuma, Naoko Sakuma and Shinobu Masaki, 'Sex Differences in Cognitive Abilities: A Cross-Cultural Perspective', *Neuropsychologia*, Vol. 28, No. 10 (1990), pp. 1063–77.

24 **no female edge on the verbal section of the SAT.** Benbow, 'Sex Differences in Mathematical Reasoning Ability', *op. cit.*

has been designed consciously to minimize sex differences. Diane Alington and Russell Leaf, 'Elimination of SAT-Verbal Sex Differences Was Due to Policy-Guided Changes in Item Content', *Psychological Reports*, Vol. 68, No. 2 (1991), pp. 541–2.

Nearly one-fourth of the verbal portion. See *Taking the SAT, 1992–1993*, Educational Testing Service, Princeton, New Jersey, 1992.

On standardized tests that evaluate language skills more extensively... Halpern, *op. cit.*, p. 92.

25 **Among stutterers, males outnumber females by three or four to one.** Halpern, *ibid.*, p. 65.

three times as many boys as girls with severe dyslexia. Joan Finucci and Barton Chiles, 'Are There Really More Dyslexic Boys Than Girls?' in Alice Ansara, Norman Geschwind and Albert Galaburda (eds), *Sex Differences in Dyslexia*, The Orton Dyslexia Society, Towson, Maryland, 1981, pp. 1–10.

27 **Bever's maze experiments.** As this is written, Bever is putting together a manuscript that describes much of his work, but to date little of it has been published. Some of it appears in Thomas Bever, 'The Logical and Extrinsic Sources of Modularity' in M. Gunnar and M. Maratsos (eds), *Modularity and Constraints in Language and Cognition*, Lawrence Erlbaum and Associates, Hillsdale, New Jersey, 1992.

28   **study by Liisa Galea and Doreen Kimura.** Liisa A. M. Galea and Doreen Kimura, 'Sex Differences in Route Learning', *Personality and Individual Differences*, Vol. 14, No. 1 (1993), pp. 53–65.
29   **study by Diane McGuinness and Janet Sparks.** Diane McGuinness and Janet Sparks, 'Cognitive Style and Cognitive Maps: Sex Differences in Representations of a Familiar Terrain', *Journal of Mental Imagery*, Vol. 7 (1983), pp. 91–100.
     study by Leon Miller and Viana Santoni. Leon K. Miller and Viana Santoni, 'Sex Differences in Spatial Abilities: Strategic and Experiential Correlates', *Acta Psychologia*, Vol. 62 (1986), pp. 225–35.
30   **In a similar study done the same year.** Shawn L. Ward, Nora Newcombe and Willis F. Overton, 'Turn Left at the Church, or Three Miles North: A Study of Direction Giving and Sex Differences', *Environment and Behavior*, Vol. 18, No. 2 (1986), pp. 192–213.
     Thomas Bever's study on language learning. Bever, 'The Logical and Extrinsic Sources of Modularity', *op. cit.*
31   **a difference in how men and women approach a task.** See, for example, Diane McGuinness, 'Sex Differences in Cognitive Style: Implications for Math Performance and Achievement' in Louis A. Penner, George M. Batsche, Howard M. Knoff, Douglas L. Nelson and Charles D. Spielberger (eds), *The Challenge of Math and Science Education: Psychology's Response*, APA Books, Washington, DC, 1993.
32   **Sex difference in town planning.** McGuinness and Sparks, *op. cit.*
33   **Women are much better at reading emotions in faces.** See, for instance, Chapter 2 in Judith A. Hall, *Nonverbal Sex Differences: Communication Accuracy and Expressive Style*, The Johns Hopkins University Press, Baltimore, 1984.
34   **Try this question on your friends.** I first ran across this question in Anne Moir and David Jessel, *Brain Sex: The Real Difference Between Men and Women*, Carol Publishing Group, New York, 1991.
     Sociologist Alice Rossi makes an important distinction. Alice S. Rossi, 'A Biosocial Perspective on Parenting', *Daedalus*, Vol. 106, No. 2 (1977), pp. 1–31.

CHAPTER 2 – A Tale of Two Sexes
page
39   **the women's events included at the Olympic Games.** Uriel Simri, *Women at the Olympic Games*, Wingate Institute for Physical Education and Sport, Netanya, Israel, 1979.
40   **Men are about nine percent taller than women.** National Center for Health Statistics, 'Anthropometric Reference Data and Prevalence of Overweight: United States, 1976–1980', *Vital & Health Statistics*, Series 11, No. 238 (1987), pp. 23–5.
     males are more muscular than females. Diane McGuinness, 'Sensorimotor Biases in Cognitive Development', in Roberta L. Hall *et al.* (eds), *Male–Female Differences: A Bio-Cultural Perspective*, Praeger, New York, 1985, pp. 57–126.
     how sensitive males and females are to loud sounds. Diane

McGuinness, 'Hearing: Individual Differences in Perceiving', *Perception*, Vol. 1 (1972), pp. 465–73.

another researcher found just as large a sex difference. Colin D. Elliott, 'Noise Tolerance and Extraversion in Children', *British Journal of Psychology*, Vol. 62 (1971), pp. 375–80.

41 **men have slightly better 'visual acuity'.** Diane McGuinness, 'Away From A Unisex Psychology: Individual Differences in Visual Perception', *Perception*, Vol. 5 (1976), pp. 279–94.

this difference becomes bigger when the target is moving. Albert Burg, 'Visual Acuity as Measured by Dynamic and Spatial Tests: A Comparative Evaluation', *Journal of Applied Psychology*, Vol. 50 (1966), pp. 460–6.

people with better 'dynamic visual acuity' had fewer accidents. Albert Burg, 'Vision and Driving: A Report on Research', *Human Factors*, Vol. 13, No. 1 (1971), pp. 79–87.

women are involved in more accidents per mile driven. National Safety Council, *Accident Facts, 1992 Edition*, Itasca, Illinois, 1992, p. 59.

men are sixty percent more likely to be involved in fatal accidents. National Safety Council, *ibid.*

In the dark, the sex difference in vision is reversed. McGuinness, 'Away From a Unisex Psychology', *op. cit.*

women have better peripheral vision. Lesley Barnes Brabyn and Diane McGuinness, 'Gender Differences in Response to Spatial Frequency and Stimulus Orientation', *Perception and Psychophysics*, Vol. 26 (1979), pp. 319–24.

42 **females more sensitive to touch.** Sidney Weinstein, 'Intensive and Extensive Aspects of Tactile Sensitivity as a Function of Body Part, Sex and Laterality' in Dan R. Kenshalo, *The Skin Senses*, Charles C. Thomas, Springfield, Illinois, 1968, pp. 195–222.

women can detect fainter odors. Richard L. Doty, 'Gender and Endocrine-Related Influences on Human Olfactory Perception' in Herbert L. Meiselman and Richard S. Rivlin, *Clinical Measurement of Taste and Smell*, Macmillan Publishing Company, New York, 1986, pp. 377–413.

women are better at recognizing smells. Richard L. Doty, Paul Shaman, Steven L. Applebaum, Ronite Giberson, Lenore Siksorski and Lysa Rosenberg, 'Smell Identification Ability: Changes with Age', *Science*, Vol. 226 (1984), pp. 1441–3. Richard L. Doty, Steven L. Applebaum, Hiroyuki Zusho and R. Gregg Settle, 'Sex Differences in Odor Identification Ability: A Cross-Cultural Analysis', *Neuropsychologia*, Vol. 23, No. 5 (1985), pp. 667–72.

It's the same story for taste. Richard L. Doty, 'Gender and Reproductive State Correlates of Taste Perception in Humans' in T. E. McGill, D. A. Dewsbury and B. D. Sachs (eds), *Sex and Behavior: Status and Prospectus*, Plenum Press, New York, 1978, pp. 337–62. Also see Mary Anne Baker, 'Sensory Functioning' in Mary Anne Baker (ed.), *Sex Differences in Human Performance*, John Wiley & Sons, New York, 1987, pp. 5–36.

0 **it's males who go for the sweet.** Marie-Odile Monneuse, France Bellisle

and Jeanine Louis-Sylvestre, 'Impact of Sex and Age on Sensory Evaluation of Sugar and Fat in Dairy Products', *Physiology and Behavior*, Vol. 50 (1991), pp. 1111–17. See also M. T. Conner and David A. Booth, 'Preferred Sweetness of a Lime Drink and Preference for Sweet Over Non-Sweet Foods, Related to Sex and Reported Age and Body Weight', *Appetite*, Vol. 10, No. 1 (1988), pp. 25–35.

43 **men and women prefer their fat in different forms.** Adam Drewnowski, Candace Kurth, Jeanne Holden-Wiltse and Jennifer Saari, 'Food Preferences in Human Obesity: Carbohydrates Versus Fats', *Appetite*, Vol. 18 (1992), pp. 207–21. Drewnowski's study actually focused only on obese subjects, but he says that the results are likely to be true for other men and women as well.

men are more likely to enjoy hot, spicy foods. Thomas R. Alley and W. Jeffrey Burroughs, 'Do Men Have Stronger Preferences for Hot, Unusual and Unfamiliar Foods?', *The Journal of General Psychology*, Vol. 118, No. 3 (1991), pp. 201–14.

Women have better manual dexterity. A number of studies have examined the female advantage in motor control. Here are three: Virgil Mathiowetz, Gloria Volland, Nancy Kashman and Karen Weber, 'Adult Norms for the Box and Block Test of Manual Dexterity', *American Journal of Occupational Therapy*, Vol. 39, No. 6 (1985), pp. 386–91; Abdur Rahim and Basu Sharma, 'Sex Differences in Performance on a Combined Manual and Decision Task', *Perceptual and Motor Skills*, Vol. 73, No. 2 (1991), pp. 695–700; and Jacqueline Agnew, Karen Bolla-Wilson, Claudia H. Kawas and Margit L. Bleecker, 'Purdue Pegboard Age and Sex Norms for People 40 Years Old and Older', *Developmental Neuropsychology*, Vol. 4, No. 1 (1988), pp. 29–35. But at least one study claims that the sex difference in manual dexterity may disappear if finger size is taken into account: Michael Peters, Philip Servos and Russell Day, 'Marked Sex Differences on a Fine Motor Skill Task Disappear When Finger Size is Used as a Covariate', *Journal of Applied Psychology*, Vol. 75, No. 1 (1990), pp. 87–90.

Kimura and Watson's throwing and 'intercepting' test. Neil V. Watson and Doreen Kimura, 'Nontrivial Sex Differences in Throwing and Intercepting: Relation to Psychometrically Defined Spatial Functions', *Personality and Individual Differences*, Vol. 12, No. 5 (1991), pp. 375–85.

44 **boys are more accurate throwers from an early age.** Diane Lunn and Doreen Kimura, 'Spatial Abilities in Preschool-Aged Children', Department of Psychology Research Bulletin No. 681, University of Western Ontario, London, Ontario, 1989.

the female advantage starts at conception. The numbers for male and female mortality rates from conception on are available in a variety of sources. See, for instance, Thomas Gualtieri and Robert E. Hicks, 'An Immunoreactive Theory of Selective Male Affliction', *The Behavioral and Brain Sciences*, Vol. 8 (1985), pp. 427–41, or David C. Taylor, 'Mechanisms of Sex Differentiation: Evidence From Disease' in J. Ghesquiere, R. D. Martin and F. Newcombe (eds), *Human Sexual Dimorphism*, Taylor and Francis, Philadelphia, 1985, pp. 169–246.

a likely explanation for why more male fetuses die. Judith M. Stillion,

Death and the Sexes, Hemisphere Publishing Corporation, Washington, 1985, p. 19.

why men die so young. For a good discussion of these factors in the death rates of men and women see Stillion, *ibid.*, pp. 15–33.

45 **a study on mice.** Naoto Kimura, Hiroyuki Yoshimura and Nobuya Ogawa, 'Sex differences in stress-induced gastric ulceration: Effects of castration and ovariectomy', *Psychobiology*, Vol. 15 No. 2 (1987), pp. 175–8.

46 **in their landmark book.** Eleanor Emmons Maccoby and Carol Nagy Jacklin, *The Psychology of Sex Differences*, Stanford University Press, Stanford, California, 1974, pp. 349–51.

47 **the real reason that people continue to believe these myths.** Maccoby and Jacklin, *ibid.*, p. 355.

48 **Boys are indeed more active than girls.** The definitive review on sex differences in activity level, encompassing one hundred and twenty-seven independent studies, is Warren O. Eaton and Lesley Reid Enns, 'Sex Differences in Human Motor Activity Level', *Psychological Bulletin*, Vol. 100, No. 1 (1986), pp. 19–28.

The female preference for dolls may reflect... For a good discussion of sex differences in parenting, see Alice S. Rossi, 'Gender and Parenthood' in Alice S. Rossi (ed.), *Gender and the Life Course*, Aldine, New York. 1985, pp. 161–91.

In 1990, McGuinness and several students... Diane McGuinness, 'Behavioral Tempo in Pre-School Boys and Girls', *Learning and Individual Differences*, Vol. 2, No. 3 (1990), pp. 315–25.

49 **In the 1970s Janet Lever...** Janet Lever, 'Sex Differences in the Games Children Play', *Social Problems*, Vol. 23 (1976), pp. 478–87.

50 **Boys' games are more complex than girls' games.** Janet Lever, 'Sex Differences in the Complexity of Children's Play and Games'. *American Sociological Review*, Vol. 43 (1978), pp. 471–83.

51 **one study of women in top management positions.** Margaret Hennig and Ann Jardim, *The Managerial Woman*, Doubleday, New York, 1977.

a trend among businesses to encourage more 'feminine' skills. See, for instance, Margo E. Garen, 'A Management Model for the '80s', *Training and Development Journal*, Vol. 36, No. 3 (1982), pp. 41–9.

52 **females have a very different psychological development.** Carol Gilligan, *In a Different Voice: Psychological Theory and Women's Development*, Harvard University Press, Cambridge, 1982.

53 **the way the two sexes speak.** Deborah Tannen, *You Just Don't Understand: Women and Men in Conversation*, William Morrow, New York, 1990.

54 **The sharpest criticism comes from other feminists.** See, for example, Katha Pollitt, 'Are Women Morally Superior to Men?', *The Nation*, 28 December 1992, pp. 799–807. Pollitt complains about the efforts of Gilligan and Tannen to picture men and women as inherently different and prefers a Marxist approach to sex differences, blaming economic inequalities for differences in how men and women behave.

ways that men and women differ in personality and temperament. These psychological differences are well known and consistent. See, for instance, James P. Kurtz and Marvin Zuckerman, 'Race and Sex Differences on the Sensation Seeking Scales', *Psychological Reports*, Vol. 43, No. 2 (1978), pp. 529–30; Harvey J. Ginsburg and Shirley M. Miller, 'Sex Differences in Children's Risk Taking Behavior', *Child Development*, Vol. 53, No. 2 (1982), pp. 426–8; Martin L. Hoffman, 'Sex Differences in Empathy and Related Behaviors', *Psychological Bulletin*, Vol. 84, No. 4 (1977), pp. 712–22; and Daniel G. Freedman and Marilyn M. DeBoer, 'Biological and Cultural Differences in Early Child Development', *Annual Review of Anthropology*, Vol. 8 (1979), pp. 579–600.

55 **sex difference in aggression.** Alice H. Eagly and Valerie J. Steffen, 'Gender and Aggressive Behavior: A Meta-Analytic Review of the Social Psychological Literature', *Psychological Bulletin*, Vol. 100, No. 3 (1986), pp. 309–30.

In many mammals... See, for example, Michael J. Meaney, Jane Stewart and William W. Beatty, 'Sex Differences in Social Play: The Socialization of Sex Roles' in Jay S. Rosenblatt, Colin Beer, Marie-Claire Busnel and Peter J. B. Slater (eds), *Advances in the Study of Behavior*, Vol. 15, Academic Press, Orlando, 1985.

the near male monopoly on violent crime. Rita J. Simon and Jean Landis, *The Crimes That Women Commit, The Punishments They Receive*, Lexington Books, Lexington, Massachusetts, 1991, particularly Chapter 4, pp. 41–56. See also Walter R. Gove, 'The Effect of Age and Gender on Deviant Behavior: A Biopsychosocial Perspective' in Alice S. Rossi, *Gender and the Life Course*, Aldine, New York, 1985.

56 **Men and women also differ in competitiveness.** Andrew Ahlgren and David W. Johnson, 'Sex Differences in Cooperative and Competitive Attitudes From the Second Through the Twelfth Grades', *Developmental Psychology*, Vol. 15, No. 2 (1979), pp. 45–9. Michael J. Strube, 'Meta-Analysis and Cross-Cultural Comparison: Sex Differences in Child Competitiveness', *Journal of Cross Cultural Psychology*, Vol. 12, No. 1 (1981), pp. 3–20.

in 'external competitiveness' the men scored much higher. Gladue had not yet written up these results as this book was written. Details come from interviews with Gladue.

a study of two hundred college students in physical education. Diane L. Gill, 'Competitiveness Among Females and Males in Physical Activity Classes', *Sex Roles*, Vol. 15 (1986), pp. 233–47.

57 **a very large sex difference in associational fluency.** Melissa Hines, 'Gonadal Hormones and Human Cognitive Development' in Jacques Balthazart (ed.), *Hormones, Brain and Behaviour in Vertebrates*, Karger, Basel (1990), pp. 51–63.

58 **Females also outperform men...** Maccoby and Jacklin, *op. cit.*, p. 351.

There is no sex difference on vocabulary tests. Janet Shibley Hyde and Marcai C. Linn, 'Gender Differences in Verbal Ability: A Meta-Analysis', *Psychological Bulletin*, Vol. 104, No. 1 (1988), pp. 53–69.

Spatial ability. See, for instance, Mark G. McGee, 'Human Spatial Abilities: Psychometric Studies and Environmental, Genetic, Hormonal and Neurological Influences', *Psychological Bulletin*, Vol. 86, No. 5 (1979), pp. 889–918.

59 **if d is around 0.2...** The definition of small, medium and large effect sizes was first suggested in J. Cohen, *Statistical Power Analysis for the Behavioral Sciences* (revised edition), Academic Press, New York, 1977.

For height, d=2.0. National Center for Health Statistics, 'Anthropometric Reference Data', *op. cit.*

the average effect size for throwing a ball. Jerry R. Thomas and Karen E. French, 'Gender Differences Across Age in Motor Performance: A Meta-Analysis', *Psychological Bulletin*, Vol. 98 (1985), pp. 260–82.

essay writing, where the effect size is only about 0.09. Hyde and Linn, 'Gender Differences in Verbal Ability', *op. cit.*

60 **an effect size of d=0.8 is typical.** See, for instance, Barbara Sanders, Mary P. Soares and Jean M. D'Aquila, 'The Sex Difference on One Test of Spatial Visualization: A Nontrivial Difference'. *Child Development*, Vol. 53 (1982), pp. 1106–10.

To put it another way... Cohen, *op. cit.*

a big male advantage on a primitive type of video tennis game. Charles S. Rebert, 'Sex Differences in Complex Visuomotor Coordination', *The Behavioral and Brain Sciences*, Vol. 3 (1980), pp. 246–7.

61 **More recently, Douglas Jackson...** As this is written, Jackson has not published the details on the sex difference in this computer game. He described it to me during a visit to the University of Western Ontario and also in subsequent phone interviews.

In shooting competitions with a moving target... The National Rifle Association keeps all sorts of records of shooting competitions, and consistently the women's records in small-bore rifle are nearly as high as the men's records. (Women don't do as well in pistol competitions as men, apparently because they don't have the arm strength to hold the pistol steady.) For some events, the holder of the open record in rifle shooting is a woman. But in trap-shooting or in events with a 'running target', the women's records are much less competitive.

high school students with high spatial ability... McGee, *op. cit.*

the sex difference in math may be mostly a result... See, for instance, Sarah A. Burnett, David M. Lane and Lewis M. Dratt, 'Spatial Visualization and Sex Differences in Quantitative Ability', *Intelligence*, Vol. 3 (1979), pp. 345–54.

62 **Besides verbal, spatial and mathematical abilities...** These tests are described in Ruth B. Ekstrom, J. W. French, H. H. Harman and D. Dermen, Kit of Factor-Referenced Cognitive Tests, Educational Testing Service, Princeton, New Jersey.

A study by Diane McGuinness and colleagues. Diane McGuinness, Amy Olson and Julia Chapman, 'Sex Differences in Incidental Recall for Words and Pictures', *Learning and Individual Differences*, Vol. 2, No. 3 (1990), pp. 263–85.

63 **In one cross-cultural study...** Virginia A. Mann, Sumiko Sasanuma, Naoko Sakuma and Shinobu Masaki, 'Sex Differences in Cognitive Abilities: A Cross-Cultural Perspective', *Neuropsychologia*, Vol. 28, No. 10 (1990), pp. 1063–77.

similar sex differences appear in other cultures. See, for instance, Max Lummis and Harold W. Stevenson, 'Gender Differences in Beliefs and Achievements: A Cross-Cultural Study', *Developmental Psychology*, Vol. 26, No. 2, (1990) pp. 254–63.
Arthur Jensen analyzed how the scores of men and women differed. Arthur R. Jensen, *Bias in Mental Testing*, The Free Press, New York, 1980, p. 624.

64 **He saw no difference between men and women on g.** As this book goes to press, Jensen's calculation that men and women have equal g has been challenged by a carefully done study on 183,000 German males and females on tests used to decide application to medical school. Douglas Jackson at the University of Western Ontario and Heinrich Stumpf at Johns Hopkins University find that men have a slightly higher g than women, with an effect size of about $d=0.3$. They also found differences in perceptual speed and memory favoring women, with effect sizes of about 0.3 and 0.6. Since the finding on g would imply that men have an IQ that averages four to five points higher than women, this is bound to be a hotly disputed finding, and it is not possible to say how the debate will turn out. Jackson is a highly respected psychometrician (a scientist who studies the measurement of intelligence) and test designer, and on the surface his study seems impressive and difficult to refute. Still, since opponents of his conclusions – and there will surely be many – have not yet had a chance to examine his work and look for flaws, it's too soon to say more. The sex difference that Jackson claims to see is small, but even the existence of such a small difference in IQ is completely unexpected. The data were presented at the meeting of the International Society for the Study of Individual Differences, 17–21 July 1993, Baltimore.

65 **more males than females are mentally retarded.** Robert G. Lehrke, 'Sex Linkage: A Biological Basis for Greater Male Variability' in R. Travis Osborne, Clyde E. Noble and Nathaniel Weyl (eds), *Human Variation: The Biopsychology of Age Race and Sex*, Academic Press, New York, 1978, pp. 171–98.
Camilla Benbow finds... David Lubinski and Camilla Persson Benbow, 'Gender Differences in Abilities and Preferences Among the Gifted: Implications for the Math-Science Pipeline', *Current Directions in Psychological Science*, Vol. 1, No. 2 (1992), pp. 61–6.
And in a massive study... David Lubinski and René V. Dawis, 'Aptitudes, Skills and Proficiencies' in M. D. Dunette and L. M. Hough (eds), *The Handbook of Industrial/Organizational Psychology*, Consulting Psychologists Press, Palo Alto, 1992, pp. 1–59.
Carol Gilligan argues... Gilligan, *op. cit.*
Deborah Tannen concludes... Tannen, *op. cit.*, p. 47.

66 **Diane McGuinness says it particularly well.** Diane McGuinness, *When Children Don't Learn: Understanding the Biology and Psychology of Learning Disabilities*, Basic Books, New York, 1985, p. 73.

CHAPTER 3 – Beyond the Birds and the Bees
page
67 **The story of Maria Patiño.** Many of the details here are drawn from Alison Carlson, 'When Is a Woman Not a Woman?', *Women's Sports & Fitness*, March 1991, pp. 24–9.
69 **the small Y chromosome ... seems to do little.** For details on the Y chromosome, see Simon Foote, Douglas Vollrath, Adrienne Hilton and David C. Page, 'The Human Y Chromosome: Overlapping DNA Clones Spanning the Euchromatic Region', *Science*, Vol. 258 (1992), pp. 60–6.
71 **In such cases of 'androgen insensitivity'...** Terry R. Brown, 'Male Pseudohermaphroditism: Defects in Androgen-Dependent Target Tissues'. *Seminars in Reproductive Endocrinology*, Vol. 5, No. 3 (1987), pp. 243–59.
72 **by no means the only exceptions to the old rule...** For an overview of the things that can go wrong in sexual development, see the August 1987 special issue of *Seminars in Reproductive Endocrinology*, Vol. 5, No. 3 (1987).
Turner's women. See, for instance, Jennifer Downey, Anke A. Ehrhardt, Akira Morishima and Jennifer J. Bell, 'Gender Role Development in Two Clinical Syndromes: Turner Syndrome Versus Constitutional Short Stature', *Journal of the American Academy of Child and Adolescent Psychiatry*, Vol. 26, No. 4 (1987), pp. 566–73.
73 **One child in every fourteen thousand...** Songya Pang *et al.*, 'Worldwide Experience in Newborn Screening for Classical Congenital Adrenal Hyperplasia Due to 21-Hydroxylase Deficiency', *Pediatrics*, Vol. 81, No. 6 (1988), pp. 866–74.
Details on CAH. Patricia A. Donohoue and Gary Berkovitz, 'Female Pseudohermaphroditism', *Seminars in Reproductive Endocrinology*, Vol. 5, No. 3 (1987), pp. 233–41.
74 **the story of Barbara.** Jared Diamond, 'Turning a Man', *Discover*, Vol. 13, No. 6 (1992), pp. 70–7.
This 5α-reductase deficiency. Julianne Imperato-McGinley, Luis Guerrero, Teofilo Gautier and Ralph E. Peterson, 'Steroid 5α-Reductase Deficiency in Man: An Inherited Form of Male Pseudohermaphroditism', *Science*, Vol. 186 (1974), pp. 1213–15.
75 **thirty-eight cases in twenty-three interrelated families.** Julianne Imperato-McGinley, Ralph E. Peterson, Teofilo Gautier and Erasmo Sturla, 'Androgens and the Evolution of Male-Gender Identity Among Male Pseudohermaphrodites With 5α-Reductase Deficiency', *The New England Journal of Medicine*, Vol. 300, No. 22 (1979), pp. 1233–7.
77 **estrogen boosts neuron growth in brain tissue.** C. Dominque Toran-Allerand, 'Organotypic Culture of the Developing Cerebral Cortex and Hypothalamus: Relevance to Sexual Differentiation', *Psychoneuroendocrinology*, Vol. 16, Nos 1–3 (1991), pp. 7–24.
78 **one further complication.** A good, though brief discussion of the roles of testosterone and estrogen in masculinizing and possibly feminizing the brain can be found in Roger A. Gorski, 'Sexual Differentiation of the Endocrine Brain and its Control' in Marcella Motta (ed.), *Brain Endocrinology*, Second Edition, Raven Press, New York, 1991, pp. 71–104.

284 THE NEW SEXUAL REVOLUTION

79 **no evidence of a gene on a sex chromosome...** See, for instance, Diane F. Halpern, *Sex Differences in Cognitive Abilities*, Lawrence Erlbaum Associates, Publishers, Hillsdale, New Jersey, 1992, pp. 105–9.
Thomas Lagueur writes in his book... Thomas Laqueur, *Making Sex: Body and Gender from the Greeks to Freud*, Harvard University Press, Cambridge, 1991, p. 4.

80 **a girl named Marie.** Laqueur, *ibid.*, p. 7. Some of the details of this story were drawn from Stephen Jay Gould, 'The Birth of the Two-Sex World', *The New York Review of Books*, 13 June 1991, pp. 11–13.

81 **the extra steps necessary to become a male may explain...** Judith M. Stillion, *Death and the Sexes*, Hemisphere Publishing Corporation, Washington, 1985, p. 19.

CHAPTER 4 – Echoes of the Womb
page

86 **Berenbaum studied a group of two dozen 'special' girls.** Sheri A. Berenbaum and Melissa Hines, 'Early Androgens are Related to Childhood Sex-Typed Toy Preferences', *Psychological Science*, Vol. 3, No. 3 (1992), pp. 203–6.

88 **CAH females are exposed to exceptionally high levels of male hormones.** Dennis J. Carson, Akimasa Okuno, Peter A. Lee, Gail Stetten, Shailaja M. Didolkar and Claude J. Migeon, 'Amniotic Fluid Steroid Levels: Fetuses With Adrenal Hyperplasia, 46,XXY Fetuses and Normal Fetuses', *American Journal of the Diseases of Children*, Vol. 136 (1982), pp. 218–22.
As early as the 1960s... Anke A. Ehrhardt, R. Epstein and John Money, 'Fetal Androgens and Female Gender Identity in the Early-treated Adrenogenital Syndrome', *Johns Hopkins Medical Journal*, Vol. 122 (1968), pp. 160–7.
In one study, Ehrhardt and Susan Baker... Anke A. Ehrhardt and Susan W. Baker, 'Fetal Androgens, Human Central Nervous System Differentiation and Behavior Sex Differences' in R. C. Friedman, R. M. Richart and R. L. Vande Wiele (eds), *Sex Differences in Behavior*, Wiley, New York, 1974.

89 **'tomboy monkeys.'** Robert W. Goy, 'Hormonally Induced Pseudohermaphroditism and Behavior' in A. G. Motulsky and W. Lentz (eds), *Birth Defects, Proceedings of the Fourth International Conference*, Excerpta Medica, Amsterdam, 1974, pp. 155–64. See also Robert Goy and Bruce McEwen (eds), *Sexual Differentiation of the Brain*, MIT Press, Cambridge, Massachusetts, 1980, pp. 44–54.
More recently, Goy... Robert W. Goy, Fred B. Bercovitch and Mary C. McBrair, 'Behavioral Masculinization is Independent of Genital Masculinization in Prenatally Androgenized Female Rhesus Macaques', *Hormones and Behavior*, Vol. 22 (1988), pp. 552–71.

90 **CAH girls don't engage in any more play fighting than normal girls.** Sheri Berenbaum has looked to see if CAH girls engage in more rough-and-tumble play and has so far seen nothing, but it may be because their mostly female playmates aren't interested in rough play, she says. CAH girls do seem to participate in more athletic activities than their sisters, she says,

but that's not quite the same thing as rough-and-tumble play.
91 **CAH females showed significantly higher spatial ability.** Susan M. Resnick, Irving I. Gottesman, Sheri A. Berenbaum and Thomas J. Bouchard, 'Early Hormonal Influences on Cognitive Functioning in Congenital Adrenal Hyperplasia', *Developmental Psychology*, Vol. 22, No. 2 (1986), pp. 191–8.

If a CAH girl were more likely to run around... This idea was suggested in Julia Sherman, 'The Problem of Sex Differences in Space Perception and Aspects of Intellectual Functioning', *Psychological Review*, Vol. 74 (1967), pp. 290–9.

To test this idea... Resnick *et al.*, *op. cit.*

92 **about one in every fourteen thousand births.** Songya Pang *et al.*, 'Worldwide Experience in Newborn Screening for Classical Congenital Adrenal Hyperplasia Due to 21-Hydroxylase Deficiency', *Pediatrics*, Vol. 81, No. 6 (1988), pp. 866–74.

99 **The male and female rats were using very different strategies.** Christina L. Williams and Warren H. Meck, 'The Organizational Effects of Gonandal Steroids on Sexually Dimorphic Spatial Ability', *Psychoneuroendocrinology*, Vol. 16, Nos. 1–3 (1991), pp. 155–76.

injected females with a sex hormone to masculinize their brains. The hormone that Williams and Meck used in these experiments was estradiol benzoate, a synthetic estrogen. It has essentially the same effect that treatment with testosterone would, since once inside the brain testosterone is transformed into estrogen, and it is actually this 'female' hormone that is responsible for much of the masculinization of the male rat brain.

Now it was the females... Williams and Meck, 'The Organizational Effects', *op. cit.*

100 **testosterone produces a rough play behavior in juvenile rats.** Michael J. Meaney, Jane Stewart, Paule Poulin and Bruce S. McEwen, 'Sexual Differentiation of Social Play in Rat Pups is Mediated by the Neonatal Androgen-Receptor System', *Neuroendocrinology*, Vol. 37 (1983) pp. 85–90.

male hormones cause rats to be more aggressive as adults. William W. Beatty, 'Gonadal Hormones and Sex Differences in Non-Reproductive Behaviors in Rodents: Organizational and Activational Influences', *Hormones and Behavior*, Vol. 12 (1979), pp. 112–63.

101 **One of the best studies of estrogen's influence.** Jane Stewart and Diane Cygan, 'Ovarian Hormones Act Early in Development to Feminize Adult Open-Field Behavior in the Rat', *Hormones and Behavior*, Vol. 14 (1980), pp. 20–32.

the actions of sex hormones in the brain... See, for instance, Roger A. Gorski, 'Sexual Differentiation of the Endocrine Brain and its Control', *Brain Endocrinology*, Second Edition, Marcella Motta (ed.), Raven Press Ltd., New York, 1991, pp. 71–104.

102 **learning differences between male and female infant rhesus macaque monkeys.** Jocelyne Bachevalier and Corinne Hagger, 'Sex Differences in the Development of Learning Abilities in Primates', *Psychoneuroendocrinology*, Vol. 16 (1991), Nos. 1–3, pp. 177–88.

103   **some evidence that the female brain matures earlier.** David C. Taylor, 'Differential Rates of Cerebral Maturation Between Sexes and Between Hemispheres', *Lancet* (1969), pp. 140–2.

a group of interrelated males living in a small Caribbean country ... Julianne Imperato-McGinley, Ralph E. Peterson, Teofilo Gautier and Erasmo Sturla, 'Androgens and the Evolution of Male-Gender Identity Among Male Pseudohermaphrodites with 5α-Reductase Deficiency', *The New England Journal of Medicine*, Volume 300, No. 22 (1979), pp. 123–7.

104   **As far as spatial ability goes, they are 'super-feminine'.** Julianne Imperato-McGinley, Marino Pichardo, Teofilo Gautier, Daniel Voyer and M. Philip Bryden, 'Cognitive Abilities in Androgen-Insensitive Subjects: Comparison With Control Males and Females from the Same Kindred', *Clinical Endocrinology*, Vol. 34 (1991), pp. 341–7.

XY women's bodies do not respond to any male hormones. Terry R. Brown, 'Male Pseudohermaphroditism: Defects in Androgen-Dependent Target Tissues', *Seminars in Reproductive Endocrinology*, Vol. 5, No. 3 (1987), pp. 243–59.

105   **They score more poorly than other women on tests of spatial ability.** There are many studies of Turner's women that find them to have very poor spatial ability and also low mathematical skills. See, for instance, Jennifer Downey, Evan J. Elkin, Anke A. Ehrhardt, Heino Meyer-Bahlburg, *et al.*, 'Cognitive Ability and Everyday Functioning in Women with Turner Syndrome', *Journal of Learning Disabilities*, Vol. 24, No. 1 (1991), pp. 32–9, or Joanne Rovet and C. Netley, 'The Mental Rotation Task Performance of Turner Syndrome Subjects', *Behavior Genetics*, Vol. 10, No. 5 (1980), pp. 437–43.

Besides their problem with spatial skills ... Elizabeth McCauley, Thomas Kay, Joanne Ito and Robert Treder, 'The Turner Syndrome: Cognitive Deficits, Affective Discrimination and Behavior Problems', *Child Development*, Vol. 58 (1987), pp. 464–73.

106   **sensation seeking in opposite-sex pairs of twins.** Susan M. Resnick, Irving I. Gottesman and Matthew McGue, 'Sensation Seeking in Opposite-Sex Twins: An Effect of Prenatal Hormones?', *Behavior Genetics*, Vol. 23, No. 4 (1993), pp. 323–29.

spatial ability in women who were twins. Shirley Cole-Harding, Ann L. Morstad and James R. Wilson, 'Spatial Ability in Members of Opposite-Sex Twin Pairs' (Abstract of meeting report.) *Behavior Genetics*, Vol. 18 (1988), p. 710.

108   **In 1982, Daniel Hier and William Crowley reported ...** Daniel B. Hier and William F. Crowley, 'Spatial Ability in Androgen-Deficient Men', *The New England Journal of Medicine*, Vol. 306, No. 20 (1982), pp. 1202–5.

110   **a consistent pattern has emerged.** June Machover Reinisch, Mary Ziemba-Davis and Stephanie A. Sanders, 'Hormonal Contributions to Sexually Dimorphic Behavioral Development in Humans', *Psychoneuroendocrinology*, Vol. 16, Nos. 1–3 (1991), pp. 213–78.

Male children exposed to high levels ... June Machover Reinisch, 'Prenatal Exposure to Synthetic Progestins Increases Potential for Aggression in Humans', *Science*, Vol. 211 (1981), pp. 1171–3.

Girls exposed to the same hormones... John Money and Anke Ehrhardt, *Man and Woman, Boy and Girl*, Johns Hopkins University Press, Baltimore, 1972.

Even when mixed with a progesterone-like hormone... The various studies are summarized in Reinisch, Ziemba-Davis and Sanders, *op. cit.*

111 **this may be true for humans as well.** Reinisch, 'Prenatal Exposure to Synthetic Progestins', *op. cit.*

the work of Jane Stewart and others. One of the most interesting, though confusing, of these other studies is Jo-Anne K. Finegan, G. Alison Niccols and Gabriel Sitarenios, 'Relations Between Prenatal Testosterone Levels and Cognitive Abilities at Four Years', *Developmental Psychology*, Vol. 28, No. 6 (1992), pp. 1075–89. Finegan finds a variety of correlations between testosterone levels and test scores in four-year-old children, but since children this young have not been well studied, it's hard to compare her results with other work and see how it fits into the overall picture.

CHAPTER 5 – My Brain's Bigger Than Your Brain
page

115 **de Lacoste's finding on the corpus callosum.** Christine de Lacoste-Utamsing and Ralph L. Holloway, 'Sexual Dimorphism in the Human Corpus Callosum', *Science*, Vol. 216 (1982), pp. 1431–2.

Levy hypothesized... Jerre Levy, 'Cerebral Lateralization and Spatial Ability', *Behavior Genetics*, Vol. 6 (1976), pp. 171–88.

118 **the first structural sex difference in the brain of a mammal.** Geoffrey Raisman and Pauline Field, 'Sexual Dimorphism in the Preoptic Area of the Rat', *Science*, Vol. 173 (1971), pp. 731–3.

119 **a sex difference in the rat brain that wasn't subtle at all.** Roger A. Gorski, John H. Gordon, James E. Shryne and Arthur M. Southam, 'Evidence for a Morphological Sex Difference Within the Medial Preoptic Area of the Rat Brain', *Brain Research*, Vol. 148 (1978), pp. 333–46.

the size of the SDN is determined by what hormones... C. D. Jacobson, V. J. Csernus, James E. Shryne and Roger A. Gorski, 'The Influence of Gonadectomy, Androgen Exposure or a Gonadal Graft in the Neonatal Rat on the Volume of the Sexually Dimorphic Nucleus of the Preoptic Area', *Journal of Neuroscience*, Vol. 1 (1981), pp. 1142–7.

a sex difference in the size of part of a songbird's brain. F. Nottebohm and A. P. Arnold, 'Sexual Dimorphism in Vocal Control Areas of the Songbird Brain', *Science*, Vol. 194 (1976), pp. 211–13.

121 **a nucleus in the human preoptic area two and one-half times larger in men.** Dick F. Swaab and E. Fliers, 'A Sexually Dimorphic Nucleus in the Human Brain', *Science*, Vol. 228 (1985), pp. 1112–15.

she published her findings on INAH-2 and 3. Laura S. Allen, Melissa Hines, James E. Shryne and Roger A. Gorski, 'Two Sexually Dimorphic Cell Groups in the Human Brain', *The Journal of Neuroscience*, Vol. 9, No. 2 (1989), pp. 497–506.

the anterior commissure is twelve percent larger in females. Laura S. Allen and Roger A. Gorski, 'Sexual Dimorphism of the Anterior Commissure and the Massa Intermedia of the Human Brain', *The Journal of*

Comparative Neurology, Vol. 312 (1991), pp. 97–104.
122 **the massa intermedia is seventy-six percent larger in females.** Allen and Gorski, *ibid.*
the bed nucleus of the stria terminalis. Laura S. Allen and Roger A. Gorski, 'Sex Difference in the Bed Nucleus of the Stria Terminalis of the Human Brain', *The Journal of Comparative Neurology*, Vol. 302 (1990), pp. 697–706.
Allen decided to look directly at living brains with MRI. Laura S. Allen, Mark F. Richey, Yee M. Chai and Roger A. Gorski, 'Sex Differences in the Corpus Callosum of the Living Human Being', *The Journal of Neuroscience*, Vol. 11, No. 4 (1991), pp. 933–42.
123 **Some researchers have found no difference in either size or shape.** One of the best negative studies is Andrew Kertesz, Marsha Polk, Janice Howell and Sandra E. Black, 'Cerebral dominance, sex and callosal size in MRI', *Neurology*, Vol. 37 (1987), pp. 1385–8.
the isthmus is larger in women than in men. Sandra F. Witelson, 'Hand and Sex Differences in the Isthmus and Genu of the Human Corpus Callosum', *Brain*, Vol. 112 (1989), pp. 799–835.
the suprachiasmatic nucleus differs in shape between the sexes. Michel A. Hofman and Dick F. Swaab, 'Sexual Dimorphism of the Human Brain: Myth and Reality', *Experimental Clinical Endocrinology*, Vol. 98, No. 2 (1991), pp. 161–70.
LeVay replicated Allen's finding. Simon LeVay, 'A Difference in Hypothalamic Structure Between Heterosexual and Homosexual Men', *Science*, Vol. 253 (1991), pp. 1034–7.
124 **Allen points out...** Laura S. Allen and Roger A. Gorski, 'Sexual Orientation and the Size of the Anterior Commissure in the Human Brain', *Proceedings of the National Academy of Sciences*, Vol. 89 (1992), pp. 7199–202.
125 **Yahr has proved that in males the SDA controls the sex act.** Pauline Yahr and Patricia D. Finn, 'Connections of the Sexually Dimorphic Area of the Gerbil Hypothalamus: Possible Pathways for Hormonal Control of Male Sexual Behavior and Scent-Marking' in Jacques Balthazart (ed.), *Hormones, Brain and Behaviour in Vertebrates, 1. Sexual Differentiation, Neuroanatomical Aspects, Neurotransmitters and Neuropeptides*, Karger, Basel, 1990, pp. 137–47.
126 **the rat's corpus callosum is larger in males than females.** Albert S. Berrebi, Roslyn H. Fitch, Diana L. Ralphe, Julie O. Denenberg, Victor L. Friedrich Jr., and Victor H. Denenberg, 'Corpus Callosum: Region-Specific Effects of Sex, Early Experience and Age', *Brain Research*, Vol. 438 (1988), pp. 216–24.
Female rat's have more unmyelinated axons... Janice M. Juraska and John Kopcik, 'Sex and environmental influences on the size and ultrastructure of the rat corpus callosum', *Brain Research*, Vol. 450 (1988), pp. 1–8.
127 **When Hines analyzed her data...** Melissa Hines, Lee Chiu, Lou Ann McAdams, Peter M. Bentler and Jim Lipcamon, 'Cognition and the Corpus Callosum: Verbal Fluency, Visuospatial Ability and Language Lateralization Related to Midsagittal Surface Areas of Callosal Regions', *Behavioral*

Neuroscience, Vol. 106, No. 1 (1992), pp. 3–14.
129 **as Stephen Jay Gould points out...** Stephen Jay Gould, *The Mismeasure of Man*, W. W. Norton & Company, New York, 1981.
brain sizes of scientists and others. Gould, *ibid.*, pp. 92–4
Le Bon's statement. As translated by Gould, *ibid.*, pp. 104–5.
131 **Maria Montessori showed that two could play at Broca's game.** Gould, *ibid.*, p. 107.
The generally accepted figure. See, for instance, Harry J. Jerison, 'The Evolution of Biological Intelligence' in R. J. Sternberg (ed.), *Handbook of Human Intelligence*, Cambridge University Press, Cambridge, 1982, pp. 723–91.
132 **a number of modern researchers have found a similar association.** Some, however, have not. Harry Jerison, for instance, concludes that there is probably no relationship between body size and brain size if you look only at males or only at females. See Harry J. Jerison, 'The Evolution of Diversity in Brain Size' in Martin E. Hahn, Craig Jensen and Bruce C. Dudek (eds), *Development and Evolution of Brain Size: Behavioral Implications*, Academic Press, New York, 1979, p. 39.
it is possible to account for all of the sex differences. K.-C. Ho, U. Roessmann, J. V. Straumfjord and G. Monroe, 'Analysis of Brain Weight', *Archives of Pathology and Laboratory Medicine*, Vol. 104 (1980), pp. 635–45.
one recent report based on autopsy records of 1,261 people. C. Davison Ankney, 'Sex Differences in Relative Brain Size: The Mismeasure of Woman, Too?', *Intelligence*, Vol. 16 (1992), pp. 329–36.
Another study that used magnetic resonance imaging. Lee Willerman, Robert Schultz, J. Neal Rutledge and Erin D. Bigler, 'In Vivo Brain Size and Intelligence, *Intelligence*, Vol. 15 (1991), pp. 223–8.
Sandra Witelson points out... Sandra F. Witelson, 'Neural Sexual Mosaicism: Sexual Differentiation of the Human Temporo-Parietal Region for Functional Assymetry', *Psychoneuroendocrinology*, Vol. 16, Nos. 1–3, pp. 131–53.
now it's important to find the real reason for the difference. For a good general discussion of what the sex difference in brain size might and might not mean, see Michael Peters, 'Sex Differences in Human Brain Size and the General Meaning of Differences in Brain Size', *Canadian Journal of Psychology*, Vol. 45, No. 4 (1991), pp. 507–22.
a similar difference in rats. Sylvia N. M. Reid and Janice M. Juraska, 'Sex Differences in the Neuron Number in the Binocular Area of the Rat Visual Cortex', *The Journal of Comparative Neurology*, Vol. 321 (1992), pp. 448–55.
134 **a group of doctors and researchers in Iowa.** Nancy C. Andreasen, Michael Flaum, Victor Swayze, Daniel S. O'Leary, Randall Alliger, Gregg Cohen, James Ehrhardt and William T. C. Yuh, 'Intelligence and Brain Structure in Normal Individuals', *American Journal of Psychiatry*, Vol. 150, No. 1 (1993), pp. 130–4.
135 **a number of recent studies have found a weak correlation.** See, for instance, Lee Willerman, Robert Schultz, J. Neal Rutledge and Erin D.

Bigler, 'In Vivo Brain Size and Intelligence', *Intelligence*, Vol. 15 (1991), pp. 223-8.

CHAPTER 6 – Not Quite the Opposite Six

page

137 **Allen speculated that this nucleus might be involved in sexual orientation.** Laura S. Allen, Melissa Hines, James E. Shryne and Roger A. Gorski, 'Two Sexually Dimorphic Cell Groups in the Human Brain', *The Journal of Neuroscience*, Vol. 9, No. 2 (1989), pp. 497-506.

the nucleus is twice as big in heterosexual men... Simon LeVay, 'A Difference in Hypothalamic Structure Between Heterosexual and Homosexual Men', *Science*, Vol. 253 (1991), pp. 1034-7.

139 **in the United States just 2.3 percent of men...** J. O. G. Billy, K. Tanfer, W. R. Grady and D. H. Klepinger, 'The Sexual Behavior of Men in the United States', *Family Planning Perspectives*, Vol. 25, No. 2 (1993), pp. 52-60.

If you demand a strict definition of homosexuality... For a good review of a number of studies of the prevalence of homosexuality see Milton Diamond, 'Homosexuality and Bisexuality in Different Populations', *Archives of Sexual Behavior*, Vol. 22, No. 4 (1993) (scheduled to appear). Also see Dean H. Hamer, Stella Hu, Victoria L. Magnuson, Nan Hu and Angela M. L. Pattatucci, 'A Linkage Between DNA Markers on the X-Chromosome and Male Sexual Orientation', *Science*, Vol. 261 (1993) pp. 321-7.

Recent sexual surveys in Britain and France. 'AIDS and Sexual Behavior in France', *Nature*, Vol. 360 (1992), pp. 407-9, and Anne M. Johnson, Jane Wadsworth, Kaye Wellings, Sally Bradshaw and Julia Field, 'Sexual Lifestyles and HIV Risk', *Nature*, Vol. 360 (1992), pp. 410-12.

the badly flawed Rinsey Report. As researchers in the field know, Kinsey's samples were not at all representative of the general population. They included, for instance, a large number of men in prisons, where the fact that there are no women available likely skews the number of men engaging in homosexual activities. Alfred Kinsey, Wardell Pomeroy and Clyde Martin, *Sexual Behavior in the Human Male*, W. B. Saunders, Philadelphia, 1948.

homosexual rights groups have continued to use it. See, for instance, Joyce Price, 'Gays Decry Survey Disputing 10% Claim', *The Washington Times*, 16 April 1993.

140 **about one in every 12,000 males and one in every 30,000 females.** A. Bakker, P. J. M. van Kesteren, L. J. G. Gooren and P. D. Bezemer, 'The Prevalence of Transsexualism in the Netherlands', *Acta Psychiatrica Scandinavica*, Vol. 87 (1993), pp. 237-8. The Netherlands study offers probably the best available numbers for several reasons. The clinic at which the researchers work does almost all of the sex-reassignment surgery in the Netherlands, and the country's tolerance of transsexuals extends to covering the reassignment surgery under national health care and allowing transsexuals to easily change their legal status from one sex to another after the surgery.

an autobiographical tale by a British travel writer. Jan Morris,

Conundrum, Harcourt Brace Jovanovich, New York, 1974.

in a few cases where a young male lost his penis... See, for instance, Milton Diamond, 'Sexual Identity, Monozygotic Twins Reared in Discordant Sex Roles and a BBC Follow-Up', *Archives of Sexual Behavior*, Vol. 11, No. 2 (1982), pp. 181–6.

141 **the 'girl' was doing well and had accepted her sex.** John Money and Anke Ehrhardt, *Man and Woman, Boy and Girl*, Johns Hopkins University Press, Baltimore, 1972. Money's views have evolved since the 1970s, and he now sees room for both biology and socialization in determining sexual identity. See John Money, *Gay, Straight and In-Between*, Oxford University Press, Oxford, 1988.

it was widely reported as such at the time. Diamond, 'Sexual Identity' *op. cit.*, p. 182.

Julianne Imperato-McGinley reported her startling study. Julianne Imperato-McGinley, Ralph E. Peterson, Teofilo Gautier and Erasmo Sturla, 'Androgens and the Evolution of Male-Gender Identity Among Male Pseudohermaphrodites with 5α-Reductase Deficiency', *The New England Journal of Medicine*, Volume 300, No. 22 (1979), pp. 1233–7.

142 **wasn't such a success story after all.** Diamond, 'Sexual Identity', *op. cit.*

'he lives as a male and seeks females as sexual partners.' Milton Diamond, 'Some Genetic Considerations in the Development of Sexual Orientation' in Marc Haug, Richard Whalen, Claude Aron and Kathie Olsen (eds), *The Development of Sex Differences and Similarities in Behavior*, Kluwer Academic Publishers, Boston, 1993.

a mixed-up sexual identity is more than just... For a discussion of the different types of transsexuals, see Ray Blanchard, Leonard H. Clemmensen and Betty W. Steiner, 'Heterosexual and Homosexual Gender Dysphoria', *Archives of Sexual Behavior*, Vol. 16 (1987), pp. 139–152, and Ray Blanchard, 'The Classification and Labeling of Nonhomosexual Gender Dysphorias', *Archives of Sexual Behavior*, Vol. 18 (1989), pp. 315–34.

Diamond is convinced... In Diamond's 'biased interaction theory' the individual interacts with his or her environment according to the sex bias set in the womb. See Milton Diamond, 'Sexual Identity and Sex Roles' in Vern L. Bullough (ed.), *The Frontiers of Sex Research*, Prometheus Books, Buffalo, New York, 1979, pp. 39–56.

143 **The SCN is nearly twice as large in homosexual men.** Dick F. Swaab and Michel A. Hofman, 'An Enlarged Suprachiasmatic Nucleus in Homosexual Men', *Brain Research* **537** (1990), pp. 141–8.

the anterior commissure in homosexual men. Laura S. Allen and Roger A. Gorski, 'Sexual Orientation and the Size of the Anterior Commissure in the Human Brain', *Proceedings of the National Academy of Sciences*, Vol. 89 (1992), pp. 7199–202.

144 **one study of thirty CAH women.** John Money, Mark Schwartz and Viola G. Lewis, 'Adult Erotosexual Status and Fetal Hormonal Masculinization and Demasculinization: 46,XY Congenital Virilizing Adrenal Hyperplasia and 46,XY Androgen-Insensitivity Syndrome Compared', *Psychoneuroendocrinology*, Vol. 9 (1984), pp. 405–14.

145 **one reported that twenty percent of the CAH women...** Ralf W.

Dittman, Marianne E. Kappes and Michael E. Kappes, 'Sexual Behavior in Adolescent and Adult Females with Congenital Adrenal Hyperplasia', *Psychoneuroendocrinology*, Vol. 17, Nos. 2/3 (1992), pp. 153–70.

another found only about five percent of CAH women to be homosexual. Rose M. Mulaikal, Claude J. Migeon and John A. Rock, 'Fertility Rates in Female Patients with Congenital Adrenal Hyperplasia Due to 21-Hydroxylase Deficiency', *New England Journal of Medicine*, Vol. 316 (1987), pp. 178–82. See also R. W. Dittman, M. E. Kappes and M. H. Kappes, 'Sexual Behavior in Adolescent and Adult Females with Congential Adrenal Hyperplasia', *Psychoneuroendocrinology*, Vol. 17 (1992), pp. 153–70.

women exposed to DES in the womb... Anke A. Ehrhardt, Heino F. L. Meyer-Bahlburg, Laura R. Rosen, Judith F. Feldman, Norma P. Veridiano, I. Zimmerman and Bruce S. McEwen, 'Sexual Orientation After Prenatal Exposure to Exogenous Estrogen', *Archives of Sexual Behavior*, Vol. 14 (1985), pp. 57–77.

their male children were more likely to be homosexual. For an overview of how such anti-miscarriage drugs seem to have affected sexual preferences of the children whose mothers took them, see Lee Ellis and M. Ashley Ames, 'Neurohormonal Functioning and Sexual Orientation: A Theory of Homosexuality-Heterosexuality', *Psychological Bulletin*, Vol. 101, No. 2 (1987), p. 233–58.

146 **they can create homosexuality in male rats...** For a review of experiments on the effects of stress on pregnant rats see Ellis and Ames, *ibid*.

German men born during and shortly after... Gunter Dörner, T. Geier, L. Ahrens, L. Krell, G. Munx, H. Sieler, E. Kittner and H. Muller, 'Prenatal Stress and Possible Aetiogenetic Factor Homosexuality in Human Males', *Endokrinologie*, Vol. 75 (1980), pp. 365–8.

in a later study... Gunter Dörner, B. Schenk, B. Schmiedel and L. Ahrens, 'Stressful Events in Prenatal Life of Bi- and Homosexual Men', *Experimental and Clinical Endocrinology*, Vol. 81 (1983), pp. 83–7.

Other scientists have failed to reproduce Dörner's results. J. Michael Bailey, Lee Willerman and Carlton Parks, 'A Test of the Maternal Stress Theory of Human Male Homosexuality', *Archives of Sexual Behavior*, Vol. 20, No. 3 (1991), pp. 277–93.

sexual preference is shaped by the sex hormones in the womb. Ellis and Ames, *op. cit.*

147 **The first really convincing study that homosexuality is inherited.** J. Michael Bailey and Richard C. Pillard, 'A Genetic Study of Male Sexual Orientation', *Archives of General Psychiatry*, Vol. 48 (1991), pp. 1089–96.

two earlier studies. Richard C. Pillard, Jeannette Poumadere and Ruth A. Caretta, 'Is Homosexuality Familial? A Review, Some Data, and a Suggestion', *Archives of Sexual Behavior*, Vol. 10 (1981), pp. 465–75, and L. L. Heston and James Shields, 'Homosexuality in Twins: A Family Study and a Registry Study', *Archives of General Psychiatry*. Vol. 18 (1968), pp. 149–160.

A similar study on lesbians. J. Michael Bailey, Richard C. Pillard, Michael

C. Neale and Yvonne Agyei, 'Heritable Factors Influence Sexual Orientation in Women', *Archives of General Psychiatry*, Vol. 50 (1993), pp. 217–23. Bailey and Pillard's work has received support from a more recent study. Frederick L. Whitam, Milton Diamond and James Martin, 'Homosexual Orientation in Twins: A Report on 61 Pairs and Three Triplet Sets', *Archives of Sexual Behavior*, Vol. 22, No. 3 (1993), pp. 187–206.

148 **Many studies have asked male homosexuals about their behavior as children.** Much of the following information comes from the review of this area of study in Melissa Hines and Richard Green, 'Human Hormonal and Neural Correlates of Sex-Typed Behaviors', *Review of Psychiatry*, Vol. 10 (1991), pp. 536–55.

One typical study... Alan P. Bell, Martin S. Weinberg and Sue K. Hammersmith, *Sexual Preference: Its Development in Men and Women*, Indiana University Press, Bloomington, Indiana, 1981.

149 **half of the homosexual men had been thought of as sissies.** Joseph Harry, *Gay Children Grow Up: Gender. Culture and Gender Deviance*, Praeger, New York, 1982.

a fifteen-year study. Richard Green, *The 'Sissy Boy Syndrome' and the Development of Homosexuality*, Yale University Press, New Haven, Connecticut, 1987.

One study of fifty-six lesbians and forty-three heterosexual women. Marcel T. Saghir and Eli Robins, *Male and Female Homosexuality; A Comprehensive Investigation*, Williams and Wilkins, Baltimore, Maryland, 1973.

150 **aggressiveness in homosexual and heterosexual men and women.** Brian Gladue, 'Aggressive Behavioral Characteristics, Hormones and Sexual Orientation in Men and Women', *Aggressive Behavior*, Vol. 17 (1991), pp. 313–26.

Homosexual men were less stereotypically masculine... Brian A. Gladue, William W. Beatty, Jan Larson and R. Dennis Staton, 'Sexual Orientation and Spatial Ability in Men and Women', *Psychobiology*, Vol. 18, No. 1 (1990), pp. 101–8.

In 1986, Geoff Sanders and Lynda Ross-Field... Geoff Sanders and Lynda Ross-Field, 'Sexual Orientation and Visuo-Spatial Ability', *Brain and Cognition*, Vol. 5 (1986), pp. 280–90.

151 **Four years later, Gladue...** Gladue, Beatty, Larson and Staton, *op. cit.*

On the spatial tests, the heterosexual men scored higher. Cheryl McCormick and Sandra Witelson, 'A Cognitive Profile of Homosexual Men Compared to Heterosexual Men and Women', *Psychoneuroendocrinology*, Vol. 16, No. 6 (1991), pp. 459–73.

When Bailey and Gladue collaborated to test... Details of this study come from personal communication with Bailey and Gladue. As this book was being finished, they still had not written up the results and submitted them for publication.

152 **a gene that produces a bias towards homosexuality in males.** Dean H. Hamer, Stella Hu, Victoria L. Magnuson, Nan Hu and Angela M. L. Pattatucci, 'A Linkage Between DNA Markers on the X-Chromosome and Male Sexual Orientation', *Science*, Vol. 261 (1993) pp. 321–7.

294 THE NEW SEXUAL REVOLUTION

the first human behaviour tracked to its roots on the chromosomes. Previously researchers have found gene mutations that affect human behavior, one of the most recent examples being an 'aggression gene' that appears to underlie outbursts of aggressive and even violent behavior in a number of men in an extended Dutch family. (Han G. Brunner *et al.*, 'X-Linked Borderline Mental Retardation with Prominent Behavioral Disturbance: Phenotype, Genetic Localization, and Evidence for Disturbed Monoamine Metabolism', *American Journal of Human Genetics*, Vol. 52 (1993), pp. 1032–9.) But Hamer's work is the first time researchers have pinpointed the location of the genes behind a behavior that involves normal motivation and thought.

153 **the genes for it should lie on the X chromosome.** Another possible explanation for male homosexuality passing through the women in a family is simply that male homosexuals are less likely than other men to have children, so paternal transmission is naturally low. Hamer acknowledges this but says that none the less the X chromosome was the right place to start looking. If the homosexuality gene did not lie on the X, he would have had to look through 22 pairs of other chromosomes, none of which seemed a more likely candidate than the others.

CHAPTER 7 – Variations on a Theme

page
157 **Men and women use different parts of their brains...** Frank B. Wood, D. Lynn Flowers and Cecile E. Naylor, 'Cerebral Laterality in Functional Neuroimaging' in Frederick L. Kitterle (ed.), *Cerebral Laterality: Theory and Research*, Lawrence Erlbaum Associates, Hillsdale, New Jersey, 1991, pp. 103–15.

161 **the pioneering neuroscientist Donald Hebb.** For a good review of Hebb's work, see Peter Milner, 'The Mind and Donald O. Hebb', *Scientific American*, January 1993, pp. 124–9.

Both have ended up at Canadian universities... One small corner of Ontario stretching from Toronto west and south to London (less than two hours drive away) holds the world's richest collection of sex difference researchers. Not only are Kimura and Witelson there, but also Elizabeth Hampson at the University of Western Ontario, Phil Bryden at the University of Waterloo, Jo-Anne Finegan at the Hospital for Sick Children in Toronto, Irwin Silverman at York University in Toronto and many others not specifically mentioned by name in this book. The wealth of talent is mostly Donald Hebb's legacy, as many of the sex difference researchers in this region are either his students, students of his students, or else have become interested in the field thanks to his students.

'**I would not expect...**' Doreen Kimura, 'Sex Differences in the Brain', *Scientific American*, September 1992, pp. 118–25.

an opinion that bothers many people. See, for instance, the letter by Melissa Hines, 'Sex Ratios at Work' in *Scientific American*, February 1993, p. 12.

163 **a stroke patient who couldn't remember the names of fruits and vegetables.** John Hart, Rita Sloan Berndt and Alfonso Caramazza,

'Category-Specific Naming Deficit Following Cerebral Infarction', *Nature*, Vol. 316 (1985), pp. 439–40.

evidence that nouns and verbs are handled by different parts of the brain. This work of Antonio Damasio at the University of Iowa was described in *Discover*, February 1993, pp. 16–17. See also Antonio Damasio, 'Category-Related Recognition Defects as a Clue to the Neural Substrates of Knowledge', *Trends in Neuroscience*, Vol. 13, No. 3 (1990), pp. 95–8.

164 **it doesn't work exactly the same way for men and women.** Herbert Lansdell, 'The Use of Factor Scores from the Wechsler-Bellevue Scale of Intelligence in Assessing Patients with Temporal Lobe Removal', *Cortex*, Vol. 6 (1968), pp. 257–68.

men are much more likely than women to suffer speech disorders. See, for instance, Doreen Kimura, 'Are Men's and Women's Brains Really Different?' *Canadian Psychology*, Vol. 28, No. 2 (1987), pp. 133–47.

Jeannette McGlone ... offered a simple hypothesis. Jeannette McGlone, 'Sex Differences in Human Brain Asymmetry: A Critical Survey', *The Behavioral and Brain Sciences*, Vol. 3 (1980), pp. 215–63.

Jerre Levy had already speculated that this might be the case. Jerre Levy, 'Cerebral Lateralization and Spatial Ability', *Behavior Genetics*, Vol. 6 (1976), pp. 171–88.

165 **The cracks started to appear after 1980.** Jeannette McGlone, 'The Neuropsychology of Sex Differences in Human Brain Organization' in Gerald Goldstein and Ralph E. Tarter (eds), *Advances in Clinical Neuropsychology*, Vol. 3, Plenum Publishing Corporation, 1986.

166 **It's not a matter of women's language centers being more spread out.** Doreen Kimura, 'Sex Differences in Cerebral Organization for Speech and Praxic Functions', *Canadian Journal of Psychology*, Vol. 37, No. 1 (1983), pp. 19–35.

167 **the language regions in men's and women brains are situated...** Doreen Kimura, 'Are Men's and Women's Brains Really Different?', *op. cit.*

when Kimura looked at damage to the front part versus the back part. Kimura, 'Sex Differences in the Brain', *op. cit*

practical, noticeable effects in men's and women's abilities. Kimura, *ibid*.

168 **a way to test Kimura's ideas on healthy people.** Richard S. Lewis and Lois Christiansen, 'Interhemispheric Sex Differences in the Functional Representation of Language and Praxic Functions in Normal Individuals', *Brain and Cognition*, Vol. 9 (1989), pp. 238–43.

169 **knocking out individual parts of human brains and seeing what happened.** Catherine Mateer, Samuel B. Polen and George A. Ojemann, 'Sexual Variation in Cortical Localization of Naming as Determined by Stimulation Mapping', *Behavioral and Brain Sciences*, Vol. 5 (1982), pp. 310–11.

171 **differences in right-left organization.** Kimura, 'Sex Differences in the Brain', *op. cit.*

spatial ability may be more focused in the right hemisphere in men. Richard S. Lewis and N. Laura Kamptner, 'Sex Differences in Spatial Task

Performance of Patients with and without Unilateral Cerebral Lesions', *Brain and Cognition*, Vol. 6 (1987), pp. 142–152.

172 **whenever a study does find a sex difference...** See, for instance, Sandra F. Witelson, 'Neural Sexual Mosaicism: Sexual Differentiation of the Human Temporo-Parietal Region for Functional Asymmetry', *Psychoneuroendocrinology*, Vol. 16, Nos. 1–3 (1993), p. 131–53.

173 **In males the right hemisphere does more of the work in identifying shapes.** Sandra Witelson, 'Sex and the Single Hemisphere: Specialization of the Right Hemisphere for Spatial Processing', *Science*, Vol. 193 (1976), pp. 425–7.

174 **The corpus callosum was larger in left-handers.** Sandra Witelson, 'The Brain Connection: The Corpus Callosum is Larger in Left-Handers', *Science*, Vol. 229 (1985), pp. 665–8.

The difference in the size... was true only for men. Sandra F. Witelson, 'Hand and Sex Differences in the Isthmus and Genu of the Human Corpus Callosum', *Brain*, Vol. 112 (1989), pp. 799–835.

there are more males than females who are left-handed. Stanley Coren, *The Left-Hander Syndrome: The Causes and Consequences of Left-Handedness*, The Free Press, New York, 1992, p. 32.

175 **Almost all right-handers have language concentrated...** Coren, *ibid.*, p. 102.

left-handers are over-represented... For a review of these findings, see M. P. Bryden, I. C. McManus and M. B. Bulman-Fleming, 'Evaluating the Empirical Support for the Geschwind-Behan-Galaburda Model of Cerebral Lateralization', *Brain and Cognition* (in press).

high levels of testosterone slow the growth of the left hemisphere. Norman Geschwind and Albert M. Galaburda, 'Cerebral Lateralization: Biological Mechanisms, Associations and Pathology: A Hypothesis and a Program for Research, I, II & III', *Archives of Neurology*, Vol. 42 (1985), pp. 428–59, 521–52 & 634–54.

Witelson suspects that it is the other way around. Sandra F. Witelson, 'Neural Sexual Mosaicism: Sexual Differentiation of the Human Temporo-Parietal Region for Functional Assymetry', *Psychoneuroendocrinology*, Vol. 16, Nos. 1–3, p. 131–53.

176 **estrogen stimulates certain types of neuron growth.** C. Dominque Toran-Allerand, 'Organotypic Culture of the Developing Cerebral Cortex and Hypothalamus: Relevance to Sexual Differentiation', *Psychoneuroendocrinology*, Vol. 16, No. 1–3 (1991), pp. 7–24.

179 **the idea that 'masculine' is the opposite of 'feminine'.** In 1973, Anne Constantinople questioned whether masculinity and femininity should be treated as opposite poles on a single spectrum in 'Masculinity-Femininity: An Exception to a Famous Dictum?', *Psychological Bulletin*, Vol. 80, No. 5 (1973), pp. 389–407. The next year, Sandra Bem suggested treating masculinity and femininity as separate dimensions in 'The Measurement of Psychological Androgyny', *Journal of Consulting and Clinical Psychology*, Vol. 42 (1974), pp. 155–62.

CHAPTER 8 – Raging Hormones

page

182 **when Hampson and Szekely compared the scores...** Elizabeth Hampson and Christine Szekely had not yet written up these results as this book was being prepared for publication. Details come from one interview with Hampson at the University of Western Ontario and many subsequent phone calls.

184 **Testosterone levels affect a male's sexual desire.** See Barbara Sherwin, 'A Comparative Analysis of the Role of Androgen in Human Male and Female Sexual Behavior: Behavioral Specificity, Critical Thresholds, and Sensitivity', *Psychobiology*, Vol. 16, No. 4 (1988), pp. 416–25.

the hormones actually alter the microscopic structure of the brain. See, for instance, Bruce S. McEwen, 'Our Changing Ideas About Steroid Effects on an Ever-Changing Brain', *Seminars in the Neurosciences*, Vol. 3 (1991), pp. 497–507.

185 **female hormones increase nurturing behavior in lab animals.** Howard Moltz, Michael Lubin, Michael Leon and Michael Numan, 'Hormonal Induction of Maternal Behavior in the Ovariectomized Nulliparous Rat', *Physiology and Behavior*, Vol. 5 (1970), pp. 1373–7.

(The same thing works in rhesus monkeys...) Joseph W. Kemnitz, Judith R. Gibber, Katherine A. Lindsay and Stephen G. Eisele, 'Effects of Ovarian Hormones on Eating Behaviors, Body Weight and Glucoregulation in Rhesus Monkeys', *Hormones and Behavior*, Vol. 23 (1989), pp. 235–50.

Men who don't produce enough testosterone... Sherwin, 'A Comparative Analysis', *op. cit.*

put such men on a testosterone-replacement regimen... Richard R. Clopper, Mary L. Voorhess, Margaret H. MacGillivray, Peter A. Lee and Barbara Mills, 'Psychosexual Behavior in Hypopituitary Men: A Controlled Comparison of Gonadotropin and Testosterone Replacement', *Psychoneuroendocrinology*, Vol. 18, No. 2 (1993), pp. 149–61.

higher testosterone levels in women are correlated... Gerianne Alexander and Barbara B. Sherwin, 'Sex Steroids, Sexual Behavior and Selection Attention for Erotic Stimuli in Women Using Oral Contraceptives', *Psychoneuroendocrinology*, Vol. 18, No. 2 (1993), pp. 91–102.

186 **Researchers still haven't unraveled all of the factors.** One good review is Patrica Schreiner-Engel, 'Female Sexual Arousability and its Relationship to Gonadal Hormones and the Menstrual Cycle', *Dissertation Abstracts International*, Vol. 41 (1980), p. 527.

a correlation in humans between testosterone and aggressive behavior. For a review of the evidence surrounding the testosterone-aggression connection, see John Archer, 'The Influence of Testosterone on Human Aggression', *British Journal of Psychology*, Vol. 82 (1991), pp. 1–28.

body builders and other athletes who take steroid hormones. A recent well-controlled study of the effects of steroids on aggression and other mood and behavioral characteristics is Tung-Ping Su, Michael Pagliaro, Peter J. Schmidt, David Pickar, Owen Wolkowitz and David R. Rubinow, 'Neuropsychiatric Effects of Anabolic Steroids in Male Normal Volunteers',

JAMA, Vol. 269, No. 21 (1993), pp. 2760–4. See also William R. Yates, Paul Perry and Scott Murray, 'Aggression and Hostility in Anabolic Steroid Users', *Biological Psychiatry*, Vol. 31, No. 12 (1992), pp. 1232–4, and P. Y. Choi, A. C. Parrott and D. Cowan, 'High-Dose Anabolic Steroids in Strength Athletes: Effects Upon Hostility and Aggression', *Human Psychopharmacology Clinical and Experimental*, Vol. 5, No. 4 (1990), pp. 349–56.

187 **the testosterone-aggression connection is not so simple.** One complication to the testosterone-aggression connection is that males probably respond differently to testosterone because of varying sensitivity of brain tissue to androgens (male hormones). The brain's response depends on both the number of androgen receptors and their sensitivity, and both probably vary from person to person.

when two groups of monkeys fight... Robert M. Rose, Thomas P. Gordon and Irwin S. Bernstein, 'Plasma Testosterone Levels in the Male Rhesus: Influences of Sexual and Social Stimuli', *Science*, Vol. 178 (1972), pp. 643–5.

A study of members of the Harvard wrestling team. Michael Elias, 'Serum Cortisol, Testosterone and Testosterone-Binding Globulin Responses to Competitive Fighting in Human Males', *Aggressive Behavior*, Vol. 7, No. 3 (1981), pp. 215–24.

Even something as simple as winning or losing a tennis match... Allen Mazur and Theodore A. Lamb, 'Testosterone, Status and Mood in Human Males', *Hormones and Behavior*, Vol. 14 (1980), pp. 236–46.

the more testosterone a man has, the more aggressive he reports himself to be. Two such studies are Brian Gladue, 'Aggressive Behavioral Characteristics, Hormones and Sexual Orientation in Men and Women', *Aggressive Behavior*, Vol. 17 (1991), pp. 313–26, and Kerrin Christiansen and Rainer Knussman, 'Androgen Levels and Components of Aggressive Behavior in Men', *Hormones and Behavior*, Vol. 21 (1987), pp. 170–80.

a way to compare testosterone and aggression more directly. Mitch Berman, Brian Gladue and Stuart Taylor, 'The Effects of Hormones, Type A Behavior Pattern, and Provocation on Aggression in Men', *Motivation and Emotion* (to appear).

190 **'tantalizing evidence' in the scientific journals.** One example of an earlier study with tantalizing but ultimately unconvincing evidence of cognitive changes across the menstrual cycle was Donald M. Broverman, William Vogel, Edward L. Klaiber, Diane Majcher, Dorothy Shea and Valerie Paul, 'Changes in Cognitive Task Performance Across the Menstrual Cycle', *Journal of Comparative and Physiological Psychology*, Vol. 95, No. 4 (1981), pp. 646–54.

192 **Her data offered solid proof.** Elizabeth Hampson and Doreen Kimura, 'Reciprocal Effects of Hormonal Fluctuations on Human Motor and Perceptual-Spatial Skills', *Behavioral Neuroscience*, Vol. 102, No. 3 (1988), pp. 456–9.

193 **The results were the same.** Elizabeth Hampson, 'Variations in Sex-Related Cognitive Abilities Across the Menstrual Cycle', *Brain and Cognition*, Vol. 14 (1990), pp. 26–43.

a study to compare women during menstruation and the preovulatory

peak. Elizabeth Hampson, 'Estrogen-Related Variations in Human Spatial and Articulatory-Motor Skills', *Psychoneuroendocrinology*, Vol. 15, No. 2 (1990), pp. 97–111. One interesting feature of this study was by coincidence nearly forty percent of the subjects, most of whom were students, were in science-related fields. Since other researchers had shown that science students of both sexes generally score much better on spatial tests than non-science students, Hampson wondered how these subjects' spatial ability, which presumably was more highly developed than that of their peers, would respond to changes in their estrogen levels. She found that although the subjects with science-related backgrounds did outscore their peers on spatial tests in both phases of the cycle, they were not immune to estrogen's ups and down. They too scored more poorly when estrogen was high than when it was low.

More recently, Irwin Silverman... The work done by Silverman and Gaulin, which was performed with their students Krista Phillips and Christine Milberg, had not been published as this book went to press. However, Silverman and Phillips have written up some similar work done a bit earlier: Irwin Silverman and Krista Phillips, 'Effects of Estrogen Changes During the Menstrual Cycle on Spatial Performance', *Ethology and Sociobiology* (in press).

194 **how short-term memory varied between the menstrual phase and luteal phase.** Susana M. Phillips and Barbara B. Sherwin, 'Variations in Memory Function and Sex Steroid Hormones Across the Menstrual Cycle', *Psychoneuroendocrinology*, Vol. 17, No. 5 (1992), pp. 497–506.

195 **only about twenty to forty percent of women...** Phillips and Sherwin, *ibid*.

a woman who had killed her boyfriend. M. B. Rosenthal, 'Insights Into the Premenstrual Syndrome', *Physician and Patient*, April 1983, pp. 46–53.

196 **estrogen does appear to give some women an emotional and physical boost.** Hampson, 'Estrogen-Related Variations', *op. cit.*

Consistently, this is not the case. See, for example, Hampson, 'Estrogen-Related Variations', *ibid.*, or Phillips and Sherwin, 'Variations in Memory Function', *op. cit.*

women receiving estrogen-replacement therapy. Susana M. Phillips and Barbara B. Sherwin, 'Effects of Estrogen on Memory Function in Surgically Menopausal Women', *Psychoneuroendocrinology*, Vol. 17, No. 5 (1992), pp. 485–95.

197 **'were not ... associated with any impairment in daily activities.'** Phillips and Sherwin, *ibid.*

198 **Testosterone does go up and down throughout the day in men...** Alain Reinberg and Michel Lagoguey, 'Circadian and Circannual Rhythms in Sexual Activity and Plasma Hormones of Five Human Males', *Archives of Sexual Behavior*, Vol. 7, No. 1 (1978), pp. 13–30.

Kimura gave mental rotations tests to men at two points during the year. Doreen Kimura, 'Sex Differences in Cognitive Function Vary with the Season', Research Bulletin #697, Department of Psychology, University of Western Ontario, 1991.

199 **In 1990 Catherine Gouchie...** Catherine Gouchie and Doreen Kimura,

'The Relationship Between Testosterone Levels and Cognitive Ability Patterns', *Psychoneuroendocrinology*, Vol. 16, No. 4 (1991), pp. 323–34. For an earlier study with similar results, see also Valerie J. Shute, James W. Pellegrino, Lawrence Hubert and Robert W. Reynolds, 'The Relationship Between Androgen Levels and Human Spatial Abilities', *Bulletin of the Psychonomic Society*, Vol. 32, 21, No. 6 (1983), pp. 465–8.

200 **Petersen studied teenaged boys and girls.** Anne C. Petersen, 'Physical Androgyny and Cognitive Functioning in Adolescence', *Developmental Psychology*, Vol. 12 (1976), pp. 524–33.

a series of recent studies on musicians and testosterone level. Marianne Hassler, 'Creative Musical Behavior and Sex Hormones: Musical Talent and Spatial Ability in the Two Sexes', *Psychoneuroendocrinology*, Vol. 17, No. 1 (1992), pp. 55–70. Marianne Hassler and Eberhard Nieschlag, 'Salivary Testosterone and Creative Musical Behavior in Adolescent Males and Females', *Developmental Neuropsychology*, Vol. 7, No. 4 (1991), pp. 503–21. Marianne Hassler, 'Testosterone and Artistic Talents', *International Journal of Neuroscience*, Vol. 56 (1991), pp. 25–38.

201 **Estrogen's influence on such things as verbal memory may arise...** Phillips and Sherwin, 'Effects of Estrogen on Memory Function', *op. cit.*

estrogen increases the density of synapses. Elizabeth Gould, Catherine S. Woolley and Bruce S. McEwen, 'The Hippocampal Formation: Morphological Changes Induced by Thyroid, Gonadal and Adrenal Hormones', *Psychoneuroendocrinology*, Vol. 16 (1991), pp. 67–84.

and the number of dendritic spines. Catherine S. Woolley, Elizabeth Gould, Maya Frankfurt and Bruce S. McEwen, 'Naturally Occurring Fluctuation in Dendritic Spine Density on Adult Hippocampal Pyramidal Neurons', *Journal of Neuroscience*, Vol. 10 (1990), pp. 4035–9.

202 **hormone-replacement therapy ... doesn't improve their spatial skills.** Daniel B. Hier and William F. Crowley, 'Spatial Ability in Androgen-Deficient Men', *The New England Journal of Medicine*, Vol. 306, No. 20 (1982), pp. 1202–5.

That something is likely to be the proper hormone environment. Gouchie and Kimura, 'The Relation Between Testosterone Levels', *op. cit.*

CHAPTER 9 – Nature/Nurture
page

203 **differences in teachers' actions toward boys and girls.** Meredith M. Kimball, 'A New Perspective on Women's Math Achievement', *Psychological Bulletin*, Vol. 105 (1989), pp. 198–214.

206 **she demands an extremely high standard of proof.** Anne Fausto-Sterling, *Myths of Gender: Biological Theories About Men and Women*, Basic Books, New York, 1985, pp. 11–12.

208 **you can't trust the science done on such a controversial topic.** This an underlying theme of Anne Fausto-Sterling's book, *Myths of Gender*, for instance. She argues that you cannot rely on science to give straight answers on a controversial subject that touches scientists personally, and so you are justified in asking scientists to spell out their political beliefs so you can judge how much to trust their data. For the same reason, she writes, it is

reasonable to apply your own political prejudices in order to decide how much evidence you will demand before you believe a certain result. Because of her political beliefs she imposes the 'highest standards of proof' on any scientific results that claim to find biological reasons for sex differences. (Apparently her standards of proof are somewhat lower for science that agrees with her political beliefs.) See pages 9–12. Stephen Jay Gould offers up a similar theme of how ideology can color science in his attack on intelligence testing, *The Mismeasure of Man*, W. W. Norton & Company, New York, 1981.

209 **if a young child develops a cataract...** Corey S. Goodman and Carla J. Shatz, 'Developmental Mechanisms That Generate Precise Patterns of Neuronal Connectivity', *Cell*, Vol. 72 (1993), pp. 77–98. For a clear layman's description of the role of the environment in the development of the visual system, see Carla J. Shatz, 'The Developing Brain', *Scientific American*, September 1992, pp. 61–7.

210 **an interesting environment stimulated the brain.** For a review see Janice M. Juraska, 'The Structure of the Rat Cerebral Cortex: Effects of Gender and the Environment' in B. Kolb and R. C. Tees (eds), *The Cerebral Cortex of the Rat*, MIT Press, Cambridge, Massachusetts, 1990, pp. 483–505.

211 **What she found surprised everyone.** Janice M. Juraska, 'Sex Differences in Dendritic Response to Differential Experience in the Rat Visual Cortex', *Brain Research*, Vol. 295 (1984), pp. 27–34.

212 **In the corpus callosum Juraska uncovered a situation...** Janice M. Juraska and John Kopcik, 'Sex and environmental influences on the size and ultrastructure of the rat corpus callosum', *Brain Research*, Vol. 450 (1988), pp. 1–8.

In the hippocampus... Janice M. Juraska, Jonathan Fitch, C. Henderson and N. Rivers, 'Sex Differences in the Dendritic Branching of Dentate Granule Cells Following Differential Experience', *Brain Research*, Vol. 333 (1985), pp. 73–80.

When she repeated the experiment with males castrated at birth... Janice M. Juraska, John R. Kopcik, Donna L. Washburne and David L. Perry, 'Neonatal Castration of Male Rats Affects the Dendritic Response to Differential Environments in Hippocampal Dentate Granule Neurons', *Psychobiology*, Vol. 16 (1988), pp. 406–10.

213 **It begins early in childhood.** For a good review of these findings see Diane McGuinness and Karl H. Pribham, 'The Origins of Sensory Bias in the Development of Gender Differences in Perception and Cognition' in Morton Bortner (ed.), *Cognitive Growth and Development: Essays in Memory of Herbert G. Birch*, Brunner/Mazel, New York, 1978.

the women more often reported seeing people... Diane McGuinness and John Symonds, 'Sex Differences in Choice Behavior: The Object–Person Dimension', *Perception*, Vol. 6 (1977), pp. 691–4.

They 'make their own environments'. Sandra Scarr and Kathleen McCartney, 'How People Make Their Own Environments: A Theory of Genotype Environmental Effects', *Child Development*, Vol. 54 (1983), pp. 424–35. This is a thought-provoking, clearly written account of how a person's nature influences how he or she interacts with the environment. Although Scarr and McCartney do not specifically address sex differences – they are talking about how people's genes drive their interactions with their environments – almost everything they say also applies to hormonally driven differences in how the sexes react to and control their environments. Also see Anke Ehrhardt, 'A Transactional Perspective on the Development of Gender Differences' in June Machover Reinisch, Leonard A. Rosenblum and Stephanie A. Sanders (eds), *Masculinity/Femininity: Basic Perspectives*, Oxford University Press, Oxford, 1987, pp. 281–5.

215 **The females spent an average of fifty percent more time in eye contact.** Joan H. Hittelman and Robert Dickes, 'Sex Differences in Neonatal Eye Contact Time', *Merrill-Palmer Quarterly*, Vol. 25, No. 3 (1979), pp. 171–84.

the women gave more physical attention to their three-month-old sons. Michael Lewis, 'State as an Infant-Environment Interaction: An Analysis of Mother–Infant Interaction as a Function of Sex', *Merrill-Palmer Quarterly*, Vol. 18 (1972), pp. 95–121.

children who get more physical comfort from their mothers... Judith Rubenstein, 'Maternal Attentiveness and Subsequent Exploratory Behavior in the Infant', *Child Development*, Vol. 38 (1967), pp. 1089–100.

it's common among young primates to spend more time with same-sex peers. See, for instance, Kees Nieuwenhuijsen, A. Koos Slob and Jacob J. van der Werff ten Bosch, 'Gender-Related Behaviors in Group-Living Stumptail Macaques', *Psychobiology*, Vol. 16, No. 4 (1988), pp. 357–71.

216 **about ten percent seemed to prefer boys instead of other girls to play with.** Sheri Berenbaum, Elizabeth Snyder and Kimberly Ketterling, 'Dissociation of Toy Preference and Playmate Preference: Evidence From Girls With Early Androgen Exposure', Paper presented at a meeting of the Society for Research into Child Development, New Orleans, March 1993. Berenbaum makes an interesting point about these results. In the past people have argued that the reason boys and girls like to play with members of their own sex is related to play preferences – you hang around with people who like to do the same types of activities. But although the CAH girls have boy-like play preferences, most of them still preferred playing with girls. This implies to Berenbaum that playmate preference is separate from toy preference – and that the 10% of CAH girls who preferred boys as playmates may have been masculinized in a different way than the other CAH girls. Given that boys who like to play with girls – the 'sissy boys' mentioned in Chapter 6 – are more likely to be homosexual when they grow up, Berenbaum is interested in testing the sexual orientation of these CAH girls later in life.

the more spatial activities a student had engaged in... Nora Newcombe, Mary M. Bandura and Dawn G. Taylor, 'Sex Differences in Spatial Ability and Spatial Activities', *Sex Roles*, Vol. 9, No. 3 (1983), pp. 377–86.

with special training it is possible to improve children's spatial skills. For a meta-analysis of many studies, see Maryann Baenninger and Nora

Newcombe, 'The Role of Experience in Spatial Test Performance: A Meta-Analysis', *Sex Roles*, Vol. 20, Nos 5–6 (1989), pp. 327–44. See also Carol Sprafkin, Lisa A. Serbin, Carol Denier and Jane M. Connor, 'Sex-Differentiated Play: Cognitive Consequences and Early Interventions' in Marsha B. Liss (ed.), *Social and Cognitive Skills*, Academic Press, New York, 1983, pp. 167–92.

217 **the 'bent twig' hypothesis.** Julia Sherman, *Sex-Related Cognitive Differences*, Thomas, Springfield, Illinois, 1978.

218 **when her daughter registered for classes in the eighth grade...** Diane F. Halpern, *Sex Differences in Cognitive Abilities*, Lawrence Erlbaum Associates, Publishers, Hillsdale, New Jersey, 1992, p. 197.

this creates a difference in the types of skills they develop. Janet Lever, 'Sex Differences in the Games Children Play', *Social Problems*, Vol. 23 (1976), pp. 478–87, and Janet Lever, 'Sex Differences in the Complexity of Children's Play and Games', *American Sociological Review*, Vol. 43 (1978), pp. 471–83.

220 **no correlation between spatial ability and childhood activities.** Susan M. Resnick, Irving I. Gottesman, Sheri A. Berenbaum and Thomas J. Bouchard, 'Early Hormonal Influences on Cognitive Functioning in Congenital Adrenal Hyperplasia', *Developmental Psychology*, Vol. 22, No. 2 (1986), pp. 191–8.

only in a subset of women who seemed to have a biological predisposition. M. Beth Casey and Mary M. Brabeck, 'Women Who Excel on a Spatial Task: Proposed Genetic and Environmental Factors', *Brain and Cognition*, Vol. 12 (1990), pp. 73–84. In particular, Casey and Brabeck found that the women who scored high on spatial ability not only had spatial experience but were likely to be right-handed with some left-handers in their family. The two researchers were testing a theory by Marian Annett which predicts that such women – right-handers from families that have some lefties – are genetically more likely to have high spatial ability than other women. Annett hypothesizes that a genetic factor she calls 'right shift' determines whether a person is right- or left-handed and also how the person's brain is organized, influencing such things as verbal and spatial ability. The right shift factor works differently in males than in females, Annett says, because the left hemisphere matures more slowly in males. See Marian Annett, *Left, Right, Hand and Brain: The Right Shift Theory*, Hillsdale, New Jersey, 1985.

the DNA of humans and chimpanzees is more than ninety-eight percent identical. See, for instance, Jared Diamond, *The Third Chimpanzee*, HarperCollins, New York, 1992.

221 **David Goldfoot and Deborah Neff observed juvenile rhesus monkeys.** David A. Goldfoot and Deborah Neff, 'Assessment of Behavioral Sex Differences in Social Contexts' in June Machover Reinisch, Leonard A. Rosenblum and Stephanie A. Sanders (eds), *Masculinity/Femininity: Basic Perspectives*, Oxford University Press, Oxford, 1987, pp. 179–95.

one group of chimps completely exterminated a second group. Jane Goodall, 'Life and Death at Gombe', *National Geographic*, May 1979, pp. 592–621.

Among vervet monkeys, juvenile females cuddle and carry small infants. Jane B. Lancaster, 'Play-Mothering: The Relations Between Juvenile Females and Young Infants Among Free-Ranging Vervet Monkeys', *Folia Primatologica*, Vol. 15 (1971), pp. 161–82.

222 **Among year-old rhesus monkeys, females approach and touch infants.** Jennifer Lovejoy and Kim Wallen, 'Sexually Dimorphic Behavior in Group-Housed Rhesus Monkeys at One Year of Age', *Psychobiology*, Vol. 16, No. 4 (1988), pp. 348–56.

in species where the fathers actually do some fathering. Eleanor Emmons Maccoby and Carol Nagy Jacklin, *The Psychology of Sex Differences*, Stanford University Press, Stanford, California, 1974, pp. 219–20.

child care in monkeys may vary with the social circumstances. Judith Rena Gibber, 'Infant-Directed Behaviors in Male and Female Rhesus Monkeys', Unpublished doctoral dissertation, University of Wisconsin-Madison, Department of Psychology, 1981.

a number of scientists take the idea quite seriously. See, for instance, Lancaster, *op. cit.*

even in families where the parents split up the child-care duties... Alice Rossi, 'Gender and Parenthood' in Alice S. Rossi (ed.), *Gender and the Life Course*, Aldine, New York, 1985, pp. 161–91.

223 **males' nurturing behavior to the young increases with exposure.** Maccoby and Jacklin, *op. cit.*, pp. 216–18.

'Everything is fifty percent genetic.' In discussing the relative influence of genes and environment, it's important to note that the 'heritability' behavioral geneticists talk about is not the same thing as 'inheritability'. Heritability refers to how much of the variation of a certain trait in a population can be attributed to genes. Suppose the trait under consideration is the number of arms a person has. This is certainly an *inheritable* trait since our genes specify two arms. However, the few cases in the population where a person has one arm or no arms are generally due to the environment, not to some genetic mutation, so the *heritability* of the number-of-arms trait is close to zero. Also note that heritability is not a fixed number, independent of the environment. If, for instance, we were able to create a risk-free society where no one lost arms due to accidents, then the heritability of the number-of-arms trait would go up, since now most of the variation in the number of arms would be due to genes.

separated twin studies. A good introductory text on behavioral genetics is Robert Plomin, J. C. DeFries and G. E. McClearn, *Behavioral Genetics*, W. H. Freeman and Company, New York, 1990. See especially Chapter 2, 'Twin Studies'.

the case of two twins named Jim. The twins have been described in a number of places. See, for instance, Edward Chen, 'Twins Reared Apart: A Living Lab', *New York Times Magazine*, 9 December 1979. See also Daniel Seligman, *A Question of Intelligence: The IQ Debate in America*, Birch Lane Press, New York, 1992, pp. 92–103.

224 **some genetic predisposition to smoking and drinking.** Alcoholism, for instance, almost certainly has a large genetic component. See, for instance, John C. Loehlin, Lee Willerman and Joseph M. Horn, 'Human Behavior

Genetics', *Annual Review of Psychology*, Vol. 39 (1988), pp. 101–33.
even the likelihood of getting divorced is influenced by genes. Matt McGue and David T. Lykken, 'Genetic Influence on Risk of Divorce', *Psycholoaical Science*, Vol. 3, No. 6 (1992), pp. 368–73.
(Correlation is a technical term...) Technically, a correlation is not really a percentage between zero and one hundred percent but rather a number from 0 to 1.0 – and to be even more precise, correlations can also be negative, so that their full range is from −1.0 to 1.0. For our purposes, however, it's less confusing and just as accurate to talk about correlations being percentages. The purists out there can convert the numbers if they like: fifty percent into 0.5, and so on.
How much is due to shared genes and how much to shared environment? For a good overview of behavioral genetics findings, see Loehlin, Willerman and Horn, 'Human Behavior Genetics', *op. cit.*

225 **Verbal and spatial abilities are ... strongly dependent on genes.** Robert Plomin, 'Environment and Genes: Determinants of Behavior', *American Psychologist*, Vol. 44, No. 2 (1989), pp. 105–11.
Memory seems to have a smaller genetic component. Nancy L. Pedersen, Robert Plomin, J. R. Nesselroade and G. E. McClearn, 'A Quantitative Genetic Analysis of Cognitive Abilities During the Second Half of the Life Span', *Psychological Science*, Vol. 3, No. 6 (1992), pp. 346–53.
Personality traits are generally less biological. Robert Plomin, 'Environment and Genes', *op. cit.*
the heritability of masculinity ... and of femininity. Jane E. Mitchell, Laura A. Baker and Carol Nagy Jacklin, 'Masculinity and Femininity in Twin Children: Genetic and Environmental Factors', *Child Development*, Vol. 60 (1989), pp. 1475–85.
whether you're an extrovert or introvert is fully half genetic. Loehlin, Willerman and Horn, *op. cit.*
shared environmental factors contribute almost nothing. Loehlin, Willerman and Horn, *ibid.*

226 **in a ten-year follow-up...** John C. Loehlin, Joseph M. Horn and Lee Willerman, 'Modeling IQ Change: Evidence from the Texas Adoption Project', *Child Development*, Vol. 60 (1989), pp. 993–1004.
Parental influence ... is strongest in childhood. Lee Willerman at the University of Texas notes that the nature of intelligence tests may be partly to blame for the apparent change in IQ heritability over time. IQ tests given to young children depend greatly on learned knowledge since it is hard to test reasoning ability at that age, and so a child's scores are greatly influenced by how much is learned at home. Later in life, when the tests depend less on information and more on reasoning ability, the influence of the home environment fades almost completely.

227 **an article surveying one hundred and seventy-two papers.** Hugh Lytton and David M. Romney, 'Parents' Differential Socialization of Boys and Girls: A Meta-Analysis', *Psychological Bulletin*, Vol. 109, No. 2 (1991), pp. 267–96.

229 **the Semai of Malaysia.** Robert Knox Dentan, *The Semai: A Nonviolent People of Malaysia*, Holt, Rinehart and Winston, New York, 1968.

In all cultures ... the male is more violent than the female. For a careful review of the data see Steven Goldberg, *The Inevitability of Patriarchy*, William Morrow & Company, New York, 1973.

men are responsible for the lion's share of the homicides. Rita J. Simon and Jean Landis, *The Crimes Women Commit, The Punishments They Receive*, Lexington Books, Lexington, Massachusetts, 1991, pp. 41–56.

230 **Among the !Kung San...** Richard B. Lee, *The !Kung San*, Cambridge University Press, Cambridge, 1979, pp. 384.

Perhaps the best cross-cultural study of sex differences. Beatrice Whiting and Carolyn Pope Edwards, 'A Cross-Cultural Analysis of Sex Differences in the Behavior of Children Ages Three Through 11', *The Journal of Social Psychology*, Vol. 91 (1973), pp. 171–88.

231 **one recent test did compare sex differences in the United States and in Japan.** Virginia A. Mann, Sumiko Sasanuma, Naoko Sakuma and Shinobu Masaki, 'Sex Differences in Cognitive Abilities: A Cross-Cultural Perspective', *Neuropsychologia*, Vol. 28, No. 10 (1990), pp. 1063–77.

232 **the sex difference in spatial skills might be decreased.** On the other hand, it's possible that giving both boys and girls extra training in spatial ability would not decrease the sex difference. There is some evidence that both sexes get about the same degree of benefit from spatial-training exercises and improve their scores on spatial tests by about the same amount. See Maryann Baenninger and Nora Newcombe, 'The Role of Experience in Spatial Test Performancs: A Meta-Analysis', *Sex Roles*, Vol. 20, Nos. 5–6 (1989), pp. 327–44.

CHAPTER 10 – Echoes of the Past

page
237 **When Gaulin compared the maze performance by sex and by species...** Steven J. C. Gaulin and Randall W. FitzGerald, 'Sex Differences in Spatial Ability: An Evolutionary Hypothesis and Test', *The American Naturalist*, Vol. 127, No. 1 (1986), pp. 74–88.

Gaulin and FitzGerald repeated the experiment... Steven J. C. Gaulin and Randall W. FitzGerald, 'Sexual Selection for Spatial-Learning Ability', *Animal Behavior*, Vol. 37 (1989), pp. 322–31.

238 **how voles caught in the wild ... compared with lab-raised voles.** Steven J. C. Gaulin and Matt S. Wartell, 'Effects of Experience and Motivation on Symmetrical-Maze Performance in the Prairie Vole', *Journal of Comparative Psychology*, Vol. 104, No. 2 (1990), pp. 183–9. In this way, voles appear to be different from rats, who perform better on mazes if they are raised in stimulating environments than if they are kept in small cages.

a physical sex difference in the brains of voles. Lucy F. Jacobs, Steven J. C. Gaulin, David F. Sherry and Gloria E. Hoffman, 'Evolution of Spatial Cognition: Sex-Specific Patterns of Spatial Behavior Predict Hippocampal Size', *Proceedings of the National Academy of Science*, Vol. 87 (1990), pp. 6349–52.

Most of the obvious physical sex differences are related to reproduction. See, for instance, Robert L. Trivers, 'Parental Investment and Sexual Selection' in B. G. Campbell (ed.), *Sexual Selection and the*

Descent of Man: 1871–1971, Aldine, Chicago, 1972, pp. 136–79.

members of one sex vie with each other to attract the opposite sex. One of the successes of evolutionary biology has been to explain which sex will be the more competitive and which the choosier in non-monogamous species. Roughly, the more competitive sex will be the one whose potential reproductive rate is faster. In mammals this is usually the male since females have to carry and nurse the young while the male has the potential to impregnate another female quickly. But in some birds and fish in which the males take care of the young, the females have a potentially higher reproductive rate and compete for mates. See Tim H. Clutton-Brock and A. C. J. Vincent, 'Sexual Selection and the Potential Reproductive Rates of Males and Females', *Nature*, Vol. 351 (1991), pp. 58–60. For an interesting description of bird species in which females compete for males, see Jared Diamond, 'Reversal of Fortune', *Discover*, April 1992, pp. 70–5.

239 **In many mammals the competition is much more direct.** Among mammals it is nearly always males competing for females and not the other way around because a winner in the male sexual competition can mate with multiple females and thus produce more children than the losers. There's no comparable prize for female mammals.

Male and female voles likely have the same genes... Gaulin has not done the hormone manipulations on the voles to prove directly that it is testosterone during development that is responsible for the sex difference in spatial ability, but Christina Williams has proved this is true for rats, and it's likely the same factors determine spatial ability in voles.

240 **From the time that modern humans appeared...** There is still some debate over just when anatomically modern humans appeared. One hundred thousand years ago is a round number that probably isn't too far off. See, for instance, Christopher B. Stringer, 'The Emergence of Modern Humans', *Scientific American*, December 1990, pp. 98–104.

humans lived in small groups of hunter-gatherers. See, for instance, Richard B. Lee and Irven De Vorer (eds), *Man the Hunter*, Aldine, Chicago, 1968.

At eleven years of age... Jerry R. Thomas and Karen E. French, 'Gender Differences Across Age in Motor Performance: A Meta-Analysis', *Psychological Bulletin*, Vol. 98, No. 2 (1985), pp. 260–82.

over many generations males gradually honed their ability... Irwin Silverman and Marion Eals, 'Sex Differences in Spatial Abilities: Evolutionary Theory and Data' in Jerome Barkow, Leda Cosmides and John Tooby (eds), *The Adapted Mind: Evolutionary Psychology and the Generation of Culture*, Oxford University Press, New York, 1992, pp. 533–49.

241 **an aiming ability.** This aiming skill is closely related to various types of spatial ability as measured by pencil-and-paper tests. See Donald Kolakowski and Robert M. Malina, 'Spatial Ability, Throwing Accuracy and Man's Hunting Heritage', *Nature*, Vol. 251 (1974), pp. 410–12, and Rosemary Jardine and N. G. Martin, 'Spatial Ability and Throwing Accuracy', *Behavior Genetics*, Vol. 13, No. 4 (1983), pp. 331–40.

the men's paleolithic partners had their own duties. One of the best studies of a modern-day hunter-gatherer culture is Richard Borshay Lee, *The !Kung San*, Cambridge University Press, Cambridge, 1979. For details on women's role in gathering and child care, including breast-feeding, see Chapter 9, 'Men, Women and Work', and Chapter 11, 'Production and Reproduction'.

242 **several ways to test the particular type of 'spatial memory'...** Silverman and Eals, *op. cit.*

women's better perceptual speed may have arisen... Doreen Kimura, 'Are Men's and Women's Brains Really Different?', *Canadian Psychology*, Vol. 28, No. 2 (1987), pp. 133–47.

243 **the language advantage may have arisen as a way to help women recall...** Silverman and Eals, *op. cit.*

the female superiority in verbal fluency ... may be merely a by-product... Kimura, *op. cit.*

once early humans began to talk. No one knows exactly when humans began talking. Kimura's explanation for the female advantage in verbal skills makes most sense if speech developed relatively late in human history and particularly if humans had already developed some type of manual communication that depended on hand signals of one sort or another. In that case, women probably had a communications advantage because of their superiority in fine motor control, and that advantage extended into spoken language.

if you took a time machine back... Of course, there's no way to be sure that humans fifty thousand years ago had brains basically identical to those in humans today, but the fossil evidence indicates that by then our ancestors were anatomically modern – their skeletons, including their skulls, looked like those of humans today.

245 **social structure of chimpanzees.** For an intriguing account of chimpanzee communities, see Jane Goodall, 'Life and Death at Gombe', *National Geographic*, May 1979, pp. 592–621. A good brief description of chimpanzees and other primates can be found in David Macdonald (ed.), *The Encyclopedia of Mammals*, Facts on File Publications, New York, 1984.

ways to measure spatial ability in monkeys. Researchers in France trained baboons to do two-dimensional rotations. When they compared the baboons with human subjects, they found that although the humans were somewhat more accurate, the baboons actually had a much faster response time. The humans took more than twice as long, on average, as the baboons. Jacques Vauclair, Joël Fagot and William D. Hopkins, 'Rotation of Mental Images in Baboons When the Visual Input is Directed to the Left Cerebral Hemisphere', *Psychological Science*, Vol. 4, No. 2 (1993), pp. 99–103.

246 **the size difference between males and females got progressively smaller.** Henry M. McHenry, 'How Big Were Early Hominids', *Evolutionary Anthropology*, Vol. 1 (1992), pp. 15–20.

the 1,154 human societies described in the *Ethnographic Atlas*. George Peter Murdock, *Ethnographic Atlas*, World Cultures, Vol. 2, No. 4 (1986).

males are interested in younger females... David M. Buss, 'Sex Differences in Human Mate Preferences: Evolutionary Hypotheses Tested in 37 Cultures', *Behavioral and Brain Sciences*, Vol. 12 (1989), pp. 1–49. David M. Buss, Randy J. Larsen, Drew Westen and Jennifer Semmelroth, 'Sex Differences in Jealousy: Evolution, Physiology and Psychology', *Psychological Science*, Vol. 3, No. 4 (1992), pp. 251–5.

CHAPTER 11 – Where Do We Go From Here?

page
255 **boys have a very different pattern of behavior than girls.** Diane McGuinness, 'Behavioral Tempo in Pre-School Boys and Girls', *Learning and Individual Differences*, Vol. 2, No. 3 (1990). pp. 322–3.
256 **Sandra Witelson argues that...** Sandra Witelson, 'Sex and the Single Hemisphere: Specialization of the Right Hemisphere for Spatial Processing', *Science*, Vol. 193 (1976), pp. 425–7.
men and women were asked to mentally go through the alphabet... Max Coltheart, Elaine Hull and Diana Slater, 'Sex Differences in Imagery and Reading', *Nature*, Vol. 253 (1975), pp. 438–40.
ways of teaching mathematics that prevent women from losing ground. For an excellent discussion of sex differences in math learning and ways to teach boys and girls more effectively, see Diane McGuinness, 'Sex Differences in Cognitive Style: Implications for Math Performance and Achievement' in Louis A. Penner, George M. Batsche, Howard M. Knoff, Douglas L. Nelson and Charles D. Spielbergert (eds), *The Challenge of Math and Science Education: Psychology's Response*, APA Books, Washington, DC, 1993.
257 **Two Stanford University researchers studied how mothers taught...** Robert D. Hess and Terexa M. McDevitt, 'Some Cognitive Consequences of Maternal Intervention Techniques: A Longitudinal Study', *Child Development*, Vol. 55 (1984), pp. 2017–30.
girls performed more poorly in those classrooms... Penelope L. Peterson and Elizabeth Fennema, 'Effective Teaching, Student Engagement in Classroom Activities, and Sex-Related Differences in Learning Mathematics', *American Educational Research Journal*, Vol. 22 (1985), pp. 309–35.
teachers who encouraged autonomy in their students... Lena Green and Don Foster, 'Classroom Intrinsic Motivation: Effects of Scholastic Level, Teacher Orientation and Gender', *Journal of Educational Research*, Vol. 80 (1987), pp. 34–9.
males' greater autonomy in learning. For a discussion of this hypothesis and other possible explanations for the male advantage on standardized math tests, see Meredith M. Kimball, 'A New Perspective on Women's Math Achievement', *Psychological Bulletin*, Vol. 105 (1989), pp. 198–214.
259 **the leakage of women from the science 'pipeline'.** See, for instance, Joe Alper, 'The Pipeline is Leaking Women All the Way Along', *Science*, Vol. 260 (1993), pp. 409–11.
260 **an article in the September 1992** *Scientific American.* Doreen Kimura, 'Sex Differences in the Brain', *Scientific American*, September 1992, pp. 118–25.
261 **Melissa Hines sharply challenged Kimura's conclusion.** Melissa Hines,

310 THE NEW SEXUAL REVOLUTION

'Sex Ratios at Work', and Doreen Kimura's reply, *Scientific American*, February 1993, p. 12.
262 **Women outnumber men among those getting degrees in...** See, for instance, National Center for Education Statistics, *Digest of Education Statistics, 1992*, Office of Educational Research and Improvement, US Department of Education.
263 **In the first year of life, boys and girls already respond differently.** Michael Lewis, 'Infants' Responses to Facial Stimuli During the First Year of Life', *Developmental Psychology*, Vol. 1, No. 2 (1969), pp. 75–86. Michael Lewis, J. Kagan and J. Kalafat, 'Patterns of Fixation in the Young Infant', *Child Development*, Vol. 37 (1966), pp. 331–41. R. B. McCall and J. Kagan, 'Attention in the Infant: Effects of Complexity, Contour, Perimeter and Familiarity', *Child Development*, Vol. 38 (1967), pp. 932–52.

a researcher asked two- to four-year-olds to illustrate a story. Evelyn W. Goodenough, 'Interest in Persons as an Aspect of Sex Differences in the Early Years', *Genetic Psychology Monographs*, Vol. 55 (1957), pp. 287–323.
264 **the kibbutz.** The classic study of sex roles in the kibbutz, which is the source of many of the facts here, is Lionel Tiger and Joseph Shepher, *Women in the Kibbutz*, Harcourt Brace Jovanovich, New York, 1975.

the rethinking of parenting and housekeeping... Melford E. Spiro, *Gender and Culture: Kibbutz Women Revisited*, Schocken Books, New York, 1980.
266 **most occupations in the kibbutz are either 'men's work' or 'women's work'.** Tiger and Shepher, *op. cit.*, Chapter 5, pp. 75–117.

women are much more likely to get degrees for... Tiger and Shepher, *ibid.* Chapter 7, pp. 159–82.
267 **the family has evolved to become the most important social unit.** Tiger and Shepher, *ibid.* Chapter 9, pp. 206–41.

kibbutz women wear stylish clothes... Spiro, *op. cit.*, pp. 15–45.

there is true equality between the sexes. Spiro, *ibid.* pp. 46–60.
269 **both sexes on the kibbutz are not happy with the range of jobs.** Spiro, *ibid.* p. 55.
271 **the mother–child bond seems to be the basic family bond.** Alice S. Rossi, 'A Biosocial Perspective on Parenting', *Daedalus*, Vol. 106, No. 2 (1977), pp. 1–31. Also see Alice S. Rossi, 'Gender and Parenthood' in Alice S. Rossi (ed.), *Gender and the Life Course*, Aldine, New York, 1985, pp. 161–91.